CINNABAR HILLS

The Quicksilver Days of New Almaden

This attempt to re-capture the life and times in New Almaden is dedicated to the sturdy miners and their families who represented the population of Englishtown, Spanishtown and the Hacienda.

Milton Lanyon - Laurence Bulmore

First Printing, July, 1967
Second Printing, July, 1968
Library of Congress Catalog Card Number A947083

Printed by the Village Printers
Los Gatos, California

ACKNOWLEDGMENTS

The authors wish to acknowledge in token of appreciation, the contributions of the Almadeners whose impressions of life in the Cinnabar Hills motivated the presentation of this material. We are especially indebted to the following individuals for their realistic expression of the New Almaden days as was experienced in Spanishtown, Englishtown and the Hacienda.

Mr. and Mrs. James Prout, the last family to leave Englishtown; John Drew, mine superintendent; John Drew Jr.; William Bunney, blacksmith; Mrs. Elizabeth Tonkin Colliver; Mrs. Lillian Hopkins Bailey, Olympia Baker, schoolteachers; Mrs. Susan Edwards Gilman; Mr. and Mrs. Fred Trevorrow; Tillie Brohaska, musician; Mrs. Gertrude Geach Shoup; Carl Carson; Clarence Malone; Mrs. Mary Hodge Hall; Edward Willoughby; David Bulmore, foreman at reduction works; Mrs. Ada Anderson Hathaway; Mrs. Anne Buzza Bulmore; Stella Lanyon Fisher; Jane Lanyon Martin; Jack Cannon; George Smoot, teamster; Dr. J. Underwood Hall; Luis Carillo; Peter Silvas; Mrs. Emily Chavez; John Fox, miner; and Leo Sullivan, musician.

Also, we are appreciative of the opportunity to show several scenes of the ghostly silence that prevailed which were taken by John Gordon, professional photographer and Carl Nipper.

FOREWORD

Over a century has passed since the New Almaden Mines became the first mining venture in California and the first to produce quicksilver in the United States. Located in the hills called the Capitancillos, twelve miles from San Jose, in the Santa Clara Valley, a fabulous cinnabar deposit was discovered in 1845. Starting with the early days of Andres Castillero and the arrival of the Barron, Forbes Company through the operations of the Quicksilver Mining Company until 1912, the Cinnabar Hills contributed its treasure to maintain a prestige of world acclaim. As a mining camp, the prevailing environment and operations, endured a routine under the complete autonomy of private ownership. No inroads were accessible for the transient or unsolicited enterprise. The mineralized area was confined within definite boundaries of company owned property. Under these conditions, the environment of the New Almaden population rarely experienced the rough, turbulent life that was generally characteristic of the western mining camp. The settlements of Englishtown, Spanishtown and the Hacienda, expressed a stability of moral, recreational and educational standards which over the years was a permanent society, little affected by outside influences.

The purpose of this material is to present in a general way, the New Almaden days as a unique mining camp which prospered for the company and provided a favorable standard of living for the miners and their families who were a dedicated group until their hour of departure. No attempt is made to give a statistical accounting of the details and procedures that constituted the years of mining operations. This is the New Almaden life as it thrived for over a half century of bountiful years and maintained a society that was in complete harmony with its environment. The early years of New Almaden would have been unrecorded were it not for the first hand, descriptive articles published by Mrs. S. A. Downer in 1854; William V. Wells, 1863; J. Ross Browne, 1865; and Mary Halleck Foote, 1877.

Since the turn of the century, the quicksilver days of New Almaden have become legendary and except for a small group of Almadeners who were born and raised within its boundaries, life on the Hill has faded into the forgotten past. The research for this publication has been supplemented by interviews with individuals who experienced the closing years of diminishing ore. The general information was acquired over many years and its acquisition was incidental as a project for publication. While the story of New Almaden is not comparable in many respects to the typical mining camp of the West, it does warrant perpetuation of name for its historical significance as a great producer of quicksilver and worthy contributor to the Gold Rush days in California.

The most important record and contribution to the preservation of New Almaden days, is the complete coverage of the subject by photography. Very few mining camps of the West were photographed so completely as New Almaden. The greater number of pictures taken between 1880 and 1912 are the work of Robert Bulmore, mine official and amateur photographer. Some of the pictures were taken by Dr. Winn, resident doctor and C. E. Watkins, an itinerant photographer, who traveled throughout the West with horse and wagon, fully equipped for his trade.

It is chiefly on the basis of this rare collection of photographs, that the story of New Almaden is presented for its historical significance and rightful place in the annals of California's great mining days. We feel that the subject which today has become legendary, warrants presentation to the many individuals who are interested in the historical mining days of California's golden era.

Milton Lanyon

Laurence Bulmore

THE MIXED TRAIN DAILY ARRIVES AT NEW ALMADEN

The Southern Pacific Company ran a branch line, standard gauge, from San Jose to New Almaden, a distance of twelve miles. This rare photograph shows the daily freight and passenger train taken about 1887, at the end of the run. The engine number 4 was built in 1863 by Danforth Cooke and Company of Patterson, New Jersey for the San Francisco and San Jose Railroad. This branch line service began operation Nov. 16, 1886 and played an important part in serving passenger and freight service during the boom years of quicksilver mining. The depot was located about two miles from the settlement and the passengers completed the journey by stagecoach. The freight was transported to the mines by teamsters. This engine was also named "Comanche" but in 1891 it was re-numbered IIII. It continued in service until January, 1898 which at this time it was dismantled in San Francisco.

CHAPTER 1

ANDRES CASTILLERO

The Discovery of the Red Cave

When the explorer, trailblazer and trapper set foot upon the land that was to become California, he found a veritable wilderness inhabited by groups of scattered aborigines whose primitive existence was isolated from the rest of the world. Unknown to the visitors, was the great storehouse of treasure that awaited the eye of those who sought nature's secret. The future destiny of this new land would be revealed throughout the world as the forces of exploitation proceeded to unlock the treasure chest of great natural resources. Great valleys of fertile soil, timbered mountains with snow melted streams, moderate climate with ample rainfall, natural harbors along a picturesque shoreline and deposits of mineral wealth that would excite the world, was the prologue to the adventuresome years of California's future.

Over the Coastal range of timbered mountains, thirty miles from the breaking surf of the Pacific Ocean, lies the fertile valley of Santa Clara. It is contained on the eastern side by uniquely sculptured foothills of sparse vegetation. Extending to the north, the valley meets the shores of San Francisco Bay, while to the south, the eye carries to infinity over ridge after ridge of low, rolling hills.

For a long period of unrecorded time, there were small, scattered groups of Indians in the Santa Clara Valley, known as Olhones or Costanes, as titled by the exploring Spaniards. Their language was similar but unrelated to that spoken by the Mitsuns in the area of Mission San Juan Bautista. They were of average stature with long, black hair and medium dark skin. As compared to other tribes to the South, they were inoffensive, mild-mannered, inferior in intelligence and existed on a low level of primitive culture. This native did little hunting and lacked the understanding to till the soil. They established their habitation in the most advantageous locations for food and water. Their survival was wholly dependent upon the most accessible items of food that were available in nature's garden. Their most energetic activity in obtaining food was gathering acorns which, when ground in their stone utensils, produced a flour. Also, added to their diet, were certain roots, weeds. berries and selected insects. On special occasions they might experience the procurement of food by hunting or fishing. They were sun worshippers and had a vague conception of evil spirits whom they attempted to appease with the performance of grotesque gestures. Their remedy for physical affliction was an exposure to heat in a small, stone structure known as Temescal. It was built with an entry and opening in the roof and situated on the banks of a creek. As the afflicted person was confined within the walls, wood was placed against the exterior and fired. When the temperature inside became unbearable, the patient would make exit and plunge into the cool stream. This practice was considered a remedy for most of their ailments.

The Olhone was not of a creative nature and produced little excepting some basketry, stone utensils and items of bone. They lived together as a family of close associates and were restrained from any tribal loyalty or affiliation. Bows, arrows and crude spears served as their weapons which, on occasions, were used in minor encounters. Their village consisted of the simplest construction or shelter beneath the native brush. Their shelter was chiefly determined by the seasons and climate conditions. For lack of any substantial evidence, it is assumed that they were completely satisfied with their meager existence and lacked the intelligence to improve their standard of living.

In comparison to other groups along the western coast, they failed to produce any evidence indicative of imagination or resourcefulness. Any ideology or interpretation of life was conceived in the fabrication of evil spirits. Typical of many primitive groups, they challenged the conception of the evil spirits by various patterns of grotesque body painting. Their limited range of color selection was derived from natural sources. From all accounts, a customary medium of adornment was a vermilion color obtained from a secreted place in the near-by hills. By some unknown means of communication, the source of this color was familiar to many Indian groups, who traveled long distances to obtain this vivid, red coloring. In the language of these primitives, it was called Mohetka.

Secundino Robles is credited as being the first Californian to discover the source of the red rock, high on the hills above the Los Alamitos creek. He had spent considerable time investigating the area with the anticipation that the colorful hills could possibly reveal gold or silver. However, his lack of understanding of minerals and the methods necessary to identify the ore, resulted in failure to achieve any material reward. After withdrawing from the activity, he passed on an account of his experiences to Antonio Sunol and Luis Chabolla, who were acquainted with the area.

It was during the year 1824, that Sunol and Chabolla, after receiving a detailed account of Roble's experiences, decided to visit the hills of the red rock. Arriving on the slopes of the highest hill, they were impressed by the exposure of red rock formations. After collecting a variety of samples, they returned to the banks of the creek to experiment with an old method used by Mexican miners. They built a small arrastra of creek stone and proceeded to pulverize the rock. Their procedure continued for many days with the greatest expectations of finding gold or silver. However, their efforts proved futile and the prospecting venture was abandoned. Their meager knowledge of minerals and deficiency in method allowed for little incentive to continue with the experimentation. With abandonment of the project, Sunol and Chabolla left the scene, unaware, that in their many hours of pulverizing the red rock, they had been working with high-grade mercurial ore and were on the threshold of a quicksilver bonanza.

In the later years, when actual mining operation would commence at the site on the hill, an interesting discovery would be made. The location had the appearance of a cave extending but a short distance into the hill. As the workers proceeded to excavate further entry, it was discovered that they had been removing cave-in material which, when removed revealed a tunnel some one hundred feet in length. Within this dark interior, were the remains of human skeletons, stoneware and varied relics. The tunnel revealed the primitive's endeavour to obtain the vermilion color. It was a general conclusion, that the tunnel into the hill without caution for safety, resulted in a cave-in near the opening, trapping the last visitors from any means of escape.

During the Fall of 1845, Andres Castillero, a captain in the military service of Mexico, arrived at Mission Santa Clara to visit the region of Upper California. Before continuing his journey to the coastal port of Monterey, where he was to meet General Jose Castro, Castillero spent several days at the Mission with Father Jose Maria del Real. During this time, he was observant of the reddish color applied about the building and a small quantity of red rock piled in the Mission yard. A brief inspection of the rock created a certain inquisitiveness on the part of Castillero, but it was time to resume his journey and he soon departed, taking with him certain impressions that would later beckon his return.

Following the western hills to the south through the open valley, Castillero arrived in Monterey after several days of travel. This coastal outpost, under the Mexican flag, was a village of fishing boats and a small military garrison. Castillero reported to the garrison and met General Jose Castro, who had recently been appointed by the dying Jose Figueroa to succeed him as governor of the territory. At this time, General Castro was engaged in the final details of preparation for a journey to visit Captain John Sutter, patriarch of New Helvetia. He recommended that Castillero accompany the party to further observe the great country to the north.

The following morning, General Castro with a small military cortege, accompanied by Castillero, left the sandy hills of oak and cypress and headed north, where they would make a short stop, to rest, at Mission San Juan Bautista. Following an over-night visit at the Mission, the Castro party continued northward through the valley of Santa Clara and on through the eastern hills, arriving days later at Sutter's Fort. Throughout the journey, Castillero was ever observant, mile after mile, of the open landscape and the adjacent hills which, at any time might reveal a tinge of vermilion coloring.

Castillero's visit at Sutter's Fort was of short duration and his strong desire to return to Mission Santa Clara prompted his early departure. His primary objective for returning was to investigate the casual rumors he had heard concerning mineral deposits and to make a careful analysis of the red rock he had observed on his recent visit. He felt quite assured in this respect, due to his training in geology, chemistry and metallurgy at the College of Mines in Mexico City. Taking leave of General Castro, he was soon retracing his steps to the Santa Clara Valley.

After an uneventful trip from Sutter's Fort, Castillero arrived at the Mission where he was heartily welcomed by Father del Real. The purpose of his quick return was readily explained and the topic of the red rock became the focal point for discussion. Father del Real knew only that the red rock had been brought to the Mission by some of the local Indians. As to the exact locality, he was not familiar, but assured Castillero that guidance to the vicinity could be found among the local natives, who had visited the area, some fifteen miles to the south.

Before becoming involved in search of the red rock hills, Castillero vigorously proceeded with the task of breaking many samples of the available rock. In the first steps of this preliminary test, he was convinced that the rock was an ore of significant properties and possibly might contain silver or gold. However, the general characteristics reminded him of a similar type which he had observed in his native Spain, where the quicksilver mines of La Mancha, the greatest in the world, had been producing for centuries. With this speculative thought, he proceeded to break the rock into fine particles, with the assumption that he was dealing with cinnabar, the ore of mercury.

Briefly familiar with the elementary procedures in which quicksilver is extracted, Castillero became more engrossed in his project. With an offer by the Mexican government of $100,000 for the discovery of mineral riches, he was ready to apply what technical knowledge he had, to unlock the secrets of the red rock. Should he be able to reveal a discovery of mineral wealth in a strange land, it would be a contribution to Mexico as well as recognition and wealth to himself.

Castillero placed some rocks in a circular arrangement and prepared a fire within. He continued the firing until there was a deep bed of extremely hot coals. Having available several pieces of flat tile, he laid them over the coals and placed on their surface the well pulverized rock. Considerable time was given in the exposure of this material to the heat, whereby, sufficient roasting would create some reaction. Assuming that the first step of his experiment was adequate, he applied cold water to the hot coals which immediately caused a steaming condition. He then held in each hand a glass container, over the rising vapor. The experiment continued with continuous firing and roasting of many samples. The project terminated when Castillero was highly elated to discover small globular particles of a silvery nature, clinging to the inner walls. He examined them closely and pressed the elusive substance with his finger until the scattered particles united into a single unit. He was favorably impressed by the results which, in his own estimation, proved that the roasted rock was cinnabar, the bearer of mercury.

The excitable Castillero made immediate plans to venture forth on a prospecting trip in quest of the red rock hills. At an early morning hour, he left the Mission with provisions and a few Indian guides on horseback and proceeded over the tree-shaded Alameda, through Pueblo San Jose and along the verdant hills of the western valley. Twenty-one years had passed since Sunol and Chabolla had experienced their labor on the banks of the Los Alamitos creek. The uneventful years had passed and the deposits of red rock had remained undisturbed by the footsteps of man.

By noon, the Castillero party had reached an area where the hills of oak and madrone appeared to fold together, closing any means of entry. They had been following a clear, trickling stream which was turning in a westerly direction toward the hills. Following the course of the stream, they arrived in a small canyon, banked on both sides with an abundant variety of foliage. Along the banks of the stream, the heavy growth of shrubbery was twisted and distorted from the heavy run-off of water during the winter season.

Following a winding trail along the banks of the creek, the party reached an open area, closed in by a high ridge of projecting rock formations. The Indians stopped and dismounted and indicated to Castillero that they had reached the end of the journey. As they unloaded their light cargo, one of the guides informed Castillero that from here they must travel by foot. Castillero looked over the horizon of twisting gullies filled with dense growth and was exuberant in the realization that his objective could be near at hand. Contoured against the skyline, colorful promontories projected their patterns in an undulating line with the folding of one ridge into another. These, were the Cinnabar Hills.

After a short rest, the horses were secured at the camp and with the Indian guides leading the way, the party commenced their ascent into the hills. The climbing and descending journey over an unmarked course, eventually led into a large, open ravine of sparse vegetation. Rising prominently from the northern side, was the apex of the hill. On its slopes projected large formations of deep, reddish stone from which nature's elements had profusely distributed many colorful samples. There was a quality of bizarreness in this setting as Castillero surveyed the total area from the depths of the ravine.

Up the steep slopes with obstructions of broken rock and jagged formations, Castillero followed the guides to a prominent indentation which was only revealed at close range. The cave-like opening was of moderate proportions for entry and shallow in depth. The Indians conveyed by gesture, that this was the red cave and the source of the color that had been supplying the visitors for a long period of unrecorded time.

Castillero leisurely inspected the shallow opening and was amazed at the walls of brilliantly colored rock which gave the impression that the interior was a mass of solid cinnabar. He broke down a variety of samples that were placed in separate piles by his helpers. As he worked on the walls, his imaginative mind conceived that this great hill was a treasure chest of quicksilver ore. The results of his exploration seemed to be growing into gigantic proportions and he was now convinced that he was on the doorstep of a great discovery. Realizing the merits of his discovery, he became aware of the need to identify himself and make known his intentions for legal claim to the property. Gathering the bags of selected samples, the party retraced its course back to their improvised camp. As it was now late in the afternoon, Castillero and his helpers prepared camp for the night.

Early the following morning, Castillero and his guides gathered their belongings and returned over their previous course to the Mission. Father del Real was extremely interested in Castillero's account of the journey and the success of locating the mysterious source of the red rock. At various intervals during the day, Castillero and Father del Real discussed in detail, the various items pertinent for obtaining official recognition of discovery and legal claim to a potential mining site.

Throughout the Mexican territory and in accordance with the existing mining laws, the first legal right to a mineral deposit was the act of discovery. In order to establish legality to any claim, the claimant must present the facts and details before a judicial tribunal which is authorized to deal with the declarations. The details for legal registry involved several requirements which consisted of a production of the ore, a description of the locality, specific points of discovery and personal details concerning the discoverer. Also, before the claim could be officially recorded and juridical possession obtained, the claimant must appear before the proper official with a written statement pertaining to the prescribed details and also, within ninety days, show development work at the site of discovery as specified.

On November 22, 1845, Andres Castillero had made the necessary preparations for filing a statement of declaration and went to the office of Pedro Chabolla, the Alcalde in Pueblo San Jose.

Following a brief conference, Castillero presented his declaration to request title to a mineral deposit. The following statement was presented and filed:
Senor Alcalde of First Nomination:

"Andres Castillero, Captain of permanent Cavalry, and at present resident in this Department, before your notorious justification makes representation, that, having discovered a vein of silver, with a ley of gold on the rancho pertaining to Jose Reyes Berryessa, retired Sergeant of the Presidio Company of San Francisco, and wishing to work it in company, I request that in conformity with the ordinances mining, you will be pleased to fix up notices in public places of the jurisdiction, in order to make sure of my right, when the time for the juridical possession may arrive, according to the laws on the matter. I pray you to provide in conformity, in which I will receive favor and justice; admitting this on common paper, there being none of the corresponding stamp."

Pueblo of San Jose Guadalupe, November twenty-second, eighteen hundred and forty-five.

Andres Castillero

Having completed the first step of entering a mining venture by the declaration of his intentions, Castillero returned to the Santa Clara Mission. He was now quite concerned in establishing a basic plan of procedure for the commencement of his mining activity. He and Father del Real discussed, in general, the customary details involved in the development of a mining project. Castillero conceived the mining operation as a company or partnership arrangement. The mine was to be known as the "Santa Clara" and the ownership would be divided into twenty-four shares. Father Jose Maria del Real was to receive four shares; four shares would be given to Don Juan Castro and four to Secundino and Teodoro Robles. The twelve remaining shares were to be retained by Castillero.

During the time of preparation for the return to the Cinnabar Hills, Castillero, by coincidence, met William Chard in Pueblo San Jose. Chard, a carpenter and general handy-man, had recently arrived in California and had already covered some of the coastal area. He had no mining experience, but in Castillero's brevity of English, he was able to discern that the venture was worth investigating. Castillero informed him that the trip to the hills would start on the following morning and designated the time and place of meeting. Castillero then returned to the Mission and concluded his plans for the eventful days ahead.

The following morning, Castillero, with horses, equipment and supplies, accompanied by several Indian helpers, left the Mission Santa Clara and proceeded to Pueblo San Jose, where William Chard was punctual in joining the party. Having partially remembered some of the details in Castillero's account of roasting the ore, Chard, with his ingenious mind, managed to acquire some old gun barrels with which he would attempt to advance the experiment.

Castillero, with his party intact, proceeded once more over his previous route to the little undisturbed canyon where a camp of a more permanent nature was arranged near the creek. The balance of the day was spent in establishing accomodations which would prove adequate for the general activities that were to follow. During the evening, Chard was able to convey to Castillero the system he had devised for using the gun barrels. Castillero was impressed and nodded approvingly.

Soon after dawn the next morning, the little party started their winding course to the red cave. Considerable time was spent collecting a variety of samples. Castillero, with his hammer, broke off fragments in the shallow opening which were sorted and placed in bags made of hide. Not all of this colorful rock was the bearer of mercury and only by segregating the samples according to their special characteristics, could the testing be of any material value. After a good representation of the rock had been assembled, the party laden with possible riches, commenced their journey back over the rugged course to camp. Each bag of samples was placed in separate piles for identification following the testing experiment.

Chard assigned the Indian helpers to the task of gathering a good supply of well seasoned wood which was plentiful in the area. Having selected sufficient creek rocks of uniform size, he arranged a circular enclosure for the fire. During this time, Castillero had found several large flat boulders and was busily engaged in the pulverization of the rock to a fine grain. Shortly, the firing commenced and continued until it had reached a stage of maximum heat. Chard made ready the gun barrels by sealing one end tightly with moist earth. Holding them upright, Castillero funneled in the grinding to almost full capacity. The gun barrels were now placed with the major portion of each on the fire. The open end of each barrel had a short extension into a container of water. Heavy firing was maintained, and it was not long until the gun barrels began to take on the glow of their fiery bed. After the metal became completely permeated with the heat, vapor began to rise from the water. As the experiment continued, the reaction of the heated contents became more evident as the release of the mercury was rapidly vaporizing into the water where it was condensing into liquid metal at the bottom of the container. Chard was greatly excited that the first experiment had proven successful, and so the process was repeated until the various samples had undergone the roasting treatment. They were able to evaluate the quality of the different samples by this primitive testing method. Castillero was now much more adept in his method of identifying and selecting the cinnabar ore. As the days passed, the little experiment continued which involved many tortuous trips of transporting ore. However, the practice was repetitious in actual results. While they had accumulated a fair amount of mercury, the meager reward was insufficient to further its continuance. Chard's visualization of another possibility that would handle more ore and increase the production of quicksilver, was ready to take shape. The gun barrel experiment had served its purpose in divulging the secrets of quicksilver extraction.

Selecting a site on the creek bank where rocks were plentiful, Chard proceeded to build a kiln of moderate proportions. The purpose of the furnace was to experiment in roasting larger quantities of ore. The simple structure was completed, and several tests were made which failed to function as expected. Lack of understanding in the proper design and a deficiency in the construction, resulted in a furnace that was technically inadequate. Unable to solve the technical difficulties encountered in the furnace operation, Chard recommended that the impractical structure be abandoned.

While progress, in advancing his mining project, was checked by the failure of the furnace, Castillero had realized sufficient evidence to indicate that he must continue operation. The amount of quicksilver which had been extracted from the gun barrel experiment proved conclusively that the cinnabar deposit was high grade quality. In his first declaration on file with the Alcalde, he had stated his discovery as containing silver with a ley of gold. He felt that it was imperative at this time to extend the merits of his mine with a supplementary statement revealing the deposit of mercurial ore. He and Chard conferred at length regarding the next step to be taken in devising an efficient method for quicksilver reduction. Chard presented a new idea and Castillero, being receptive to any means of experimentation, encouraged his assistant to proceed with the project while he was away for a short time in Pueblo San Jose.

On December third, Castillero was again at the office of the Alcalde with another statement of declaration in which he states that in conjunction with the claim of silver with a ley of gold, he had definitely discovered quicksilver with sufficient evidence to substantiate his claim. He wished this statement to be placed on record and possession to be granted according to the law.

The days passed and on December 30th a certificate of possession was granted to Castillero by First Alcalde, Antonio Maria Pico. The following terms were stated:

"I have granted three thousand varas (yards) of land in all directions, subject to what the general ordinance of mines may direct, it being worked in company, to which I certify, the witnesses signing with me."

Witnesses —— Antonio Sunol, Antonio Maria Pico, Alcalde Jose Noriega.

Having acquired what was termed "juridical possession" of the mining property, Castillero returned to his camp on the Los Alamitos creek, greatly assured as to his legal ownership. He was surprised to find Chard busily engaged in the mechanics of assembling six large cast iron pots which he had obtained from Monterey. They were known as Whaler's try-pots and used by the fisherman for boiling or cooking purposes.

Castillero was amazed at the ingenuity of his assistant and recognized in this improvised furnace, a feasible scheme that would keep the project alive. Chard explained his plan of placing the broken ore in three pots and using the other three as covers. The arrangement would accommodate some three to four tons of ore. Three pots were supported by large boulders, allowing sufficient space beneath for adequate firing. The cover would be sealed with mud and a pipe outlet would extend into a container of water. Castillero expressed an enthusiastic approval to this new innovation and immediately engaged his Indian helpers in the vigorous activity of transporting ore.

Several days were spent in accumulating a sufficient amount of ore. After the ore had been broken down to pebble size, the pots were loaded to a reasonable capacity. The covers were sealed into place and the firing commenced. The maintenance of maximum heat was continued until, eventually, mercurial vapor was condensing in the water. The results of the first test proved successful, and the inexperienced miners were very encouraged with the outlook for their new furnace. The activity became a steady schedule of collecting and roasting ore.

The months rolled by and the small group of quicksilver producers had been functioning in a very systematic routine. The try-pots were kept consistently active and continued in the repetitious procedure to reward the efforts of the workers. As the project reached the month of August, 1846, approximately 3,000 pounds of ore had been fired with favorable results. However, although the primitive device was functioning successfully, the results were not sufficient to carry the project to the proportions of development necessary to advance the progress of the operation. Lack of finances, equipment and supplies became the immediate problem which Castillero now recognized as the next obstacle confronting his operation. William Chard, having experienced about eight months in the chores of producing quicksilver, realized that the activity had reached a climax. He informed Castillero that he would be moving on and followed by the Indian helpers, left for parts unknown.

Andres Castillero had not lost his enthusiasm for the Santa Clara mine. He made a realistic appraisal of the situation and decided that the magnitude of his mine could never be realized without financial help, proper equipment and adequate labor. The financing of these needs was not available in the immediate area so he decided to return to Mexico and seek government support. He returned to the Santa Clara Mission and explained his plans to Father del Real, who favorably agreed that his intentions to seek help in Mexico was the solution to the problem.

Within a short space of time, after leaving the Cinnabar Hills, Castillero departed from the Santa Clara Valley, little knowing that his exploits in producing quicksilver would never be resumed. He arrived in Monterey, Mexico, and proceeded to establish the legality of his claim and gain a favorable response for financial aid from the Mexican government. During his days of official conferences, little was accomplished due to the mounting conflict between the United States and Mexico. Before anything definite from an official or financial standpoint could be realized, hostilities had reached a stage of open war and Castillero's dream of returning to the red cave to develop his great mine became more remote.

Failing to achieve any success and with the obligation of active military service becoming more imminent each day, Castillero decided to relinquish his shares in the Santa Clara Mine. It was not long before a ready customer was available in the name of Barron, Forbes Company, industrial operators, who were located in Tepic, Mexico.

The story of Andres Castillero comes to an abrupt ending during the early part of 1847. While he no doubt carried fond memories of the Cinnabar Hills, the rapid change of events had reshaped his attitude and objective for his future experiences. The cave of the red rock would become but a memory with little realization of any material reward. His identity, as the discoverer of quicksilver and the promotion of the first legal mining claim in the New World, would become indistinct in the pages of recorded accomplishments. Castillero vanished into obscurity, little aware that in the near future his discovery would attain world-wide prominence and that the hills of red rock would become one of the greatest quicksilver mines in the world during its span of existence.

———————

CHAPTER 2

NUEVO ALMADEN

The Beginning of New Almaden Quicksilver Mining

ALEXANDER FORBES
Courtesy of California Pioneers Society

Alexander Forbes and Eustace Barron headed the English industrial firm that was operating a cotton mill in Tepic, Mexico. During the year 1846, they acquired the shares of Andres Castillero and General Jose Castro, which constituted two-thirds interest in the Santa Clara Mine. Whether by chance or coincidence, the Barron, Forbes Company became destined to initiate and promote the first organized mining operation in the Mexican territory of California. The remaining eight shares were held by the original members of the Castillero combine, Father Jose Maria del Real at the Santa Clara Mission and the Robles brothers, Secundino and Teodoro. Alexander Forbes, who from all appearances assumed the executive initiative for the company, was enthusiastic to speculate on the basis of his observation of the Mexican mines and their wealth of production. The simple transaction

with Castillero in a mine which had been officially recognized appeared as a golden opportunity that involved little risk. A speculative venture in mining at a time when the relations between the United States and Mexico were becoming more hostile, clearly expressed the optimism of the new owners. The speculative attitude was based chiefly on the fact that Castillero had discovered a hill of mercurial ore and had produced a small quantity of quicksilver. In this rather abstract picture, it was not evident that the days to come would be plagued with problems of conflict and that the eight remaining shares would be circulated in a course of events eventually reaching a climax of national attention.

During the period of time following the departure of Castillero and the arrival of the Barron, Forbes representation, the mine property was under the custodianship of James Forbes, having been placed there by Father Jose Maria del Real, who assumed a certain responsibility on behalf of Castillero. The situation at this time was in a dormant stage and there had been no communication from Mexico to enlighten the local shareholders of recent developments. Forbes, who was no relation to Alexander Forbes, established residence on the property and, while maintaining vigilance, was able to ascertain that the mine was a reality of more than average value. His association with adjoining landowners, Jose Reyes Berryessa and Justo Larios, revealed some pertinent facts regarding the boundary lines of the ranches and the mine property. His consistent interrogation regarding the situation and his favorable comments of the mine, created an awareness on the part of Berryessa and Larios by which they were becoming seriously concerned. As the days passed, Forbes became more perceptive that the Castillero claim lacked sufficient detail, from a legal standpoint, to establish unquestioned ownership to a specified area of 3,000 varas. He began to concentrate his efforts more objectively, with a personal viewpoint toward bigger stakes than that of a custodian.

In the early part of 1847, Alexander Forbes had formulated his plans to make contact with the

Cinnabar Hills. He dispatched Robert Walkinshaw, an employee of the firm, to establish custodianship of the property. He had recently been informed of the situation at the mine and was not receptive to the arrangement of James Forbes as representing the firm. Walkinshaw arrived in the isolated location where his unexpected arrival was not amiably received by Forbes. Greater surprise was registered when a statement was presented indicating that the bearer, Robert Walkinshaw, was to become property custodian. Forbes was reluctant to accept the order and announced his intentions to remain on the property. He refuted the legality and authority of the Barron, Forbes Company to assume ownership. And, as a representative of the minority group, he would continue to serve in his present capacity. After several days of conflict and verbal discourse, Walkinshaw obtained the services of the law in San Jose, which promptly evicted his predecessor.

Following the eviction of Forbes, Walkinshaw had a small dwelling constructed and settled down to the duties of official custodian. During the months that passed while waiting for the arrival of Alexander Forbes, Walkinshaw, whose curiosity had been aroused by the comments of Forbes, spent considerable time in covering the mine area and conferring with the bordering neighbors. The situation was apparent, in his opinion that, two points of controversy were in the making, one dealing with the boundary lines and the other which challenged legal ownership. It was not beyond his manner of reason to view the situation as open domain should future developments jeopardize the controlling interests of the Barron, Forbes Company.

In the Fall of 1847, Alexander Forbes arrived from Mexico with a sizeable crew of workers, equipment and John Young, who would superintend the initial operations. He received a general accounting from his custodian with special emphasis upon the legality of the claim. Forbes, while being receptive to the information, was not alarmed as to the development of any controversy. Within a short space of time, crews of workers were busily engaged in the preparation of facilities for operation.

On the banks of the Los Alamitos creek, at the gateway to the mine, structures were erected to serve as accommodations for the workers and also would be the location for the furnace operations. This first settlement would later be referred to as the Hacienda, taken from the Spanish term "Hacienda de Beneficio," meaning reduction works. Over several low ridges in a large, open ravine,

workers were assigned to the building of small houses which would accommodate the mining crews. This rather bizarre setting of colorful rock and native foliage would later be known as Deep Gulch, and the little settlement would become Spanishtown.

During the construction activity, a small crew of miners began the preliminary stages of prospecting and ore extraction. The site of discovery became the focal point of operation and the red cave, on the slopes of what was later to be known as Mine Hill, was subjected to vigorous treatment with picks and crowbars. In conjunction with these activities, another crew was preparing the furnace area and building retorts. These two projects would be co-ordinated by a crude but adequate road that followed the contour of the hills, joining the two areas of operation. Men and mules created a means of travel over this rugged terrain, which would serve the mining activities until 1864.

The well-planned procedure of Alexander Forbes progressed with precision and success. A marked transformation of this isolated area had been accomplished within a short space of time and the actual production of quicksilver had reached the day of commencement. During these busy days, Forbes realized the discrepancies that were not obvious in the initial transaction and the lack of sufficient detail concerning the property. While he experienced an atmosphere of conflict, he had no doubts that the Barron, Forbes Company, with controlling interest, could not maintain the initiative and authority to proceed with the operation. It was apparent that forces of opposition were methodically engaged in forcing a challenge to dispute certain obvious defects of the Castillero claim. However, with complete confidence in the state of affairs and with no critical showdown of an immediate nature, Alexander Forbes, for personal and business reasons, found it advisable to return to Mexico. A foothold had been established and his general evaluation gave every assurance that the speculative venture was a wise investment and that future days would divulge a mineral deposit of great promise. This would be a great mine, in the same category as the famous Almaden, in Spain, which had been operating for centuries. Impressed by the reputation of the fabulous Spanish mine, Forbes decided to use the same title which was derived from the Arabic, meaning "the mine." So, before the first drops of liquid silver were extracted from the roasted ore, the Santa Clara Mine of Castillero became the New Almaden Mines. The area confined within its boundaries would be known as Nuevo Almaden.

With the continuance of Robert Walkinshaw as general overseer and the work crews under capable supervision, Forbes departed for Mexico for a trip of short duration. While Walkinshaw was not intelligently informed in the details of mine operation, he managed to apply his time in surveillance and contact with the daily procedure. As the weeks passed, he became more aware of the agressive personalities that were activating measures of conflict which, subsequently, would evolve into conditions quite precarious to the welfare of the Barron, Forbes Company.

At the beginning of 1848, California was at the threshold of a transitional period. The first incident, which was to have profound influence, occurred in January with the discovery of gold by James Marshall on the American River at Coloma and soon was publicized on the streets of San Francisco by the illustrious Sam Brannan. On February 2nd the region of Upper California became the acquisition of the United States by the Treaty of Guadalupe-Hidalgo for a settlement of $18,000,000. The portals of California were open to a new frontier which would beckon a cavalcade from all walks of life by land and water. This was the prelude to the coming of the sturdy "49"ers, with their caravans of covered wagons making their course over the rugged plains and challenging the wastelands of desert heat. They would come by the thousands, each seeking reward in the land of promise. California would be reborn by this great influx of immigrants who would spread over the land and give impetus to the building of the West.

From Mariposa northward to the mountain settlement of Downieville, the fabulous Mother Lode would extend its storehouse of riches with veins of quartz and gold dust streams. Over the high timbered Sierra Nevada mountains, in the land of Washoe, Nevada, the subterranean riches of the great Comstock Lode would expend its relatively short life with the glamour and notoriety befitting a mining metropolis. The era of mining activity would spread its domain from the "richest hill on earth" in Butte, Montana, through every conceivable locale, to the notorious outpost of Tombstone, Arizona. These would be the Gold Rush days, and the exploitation of the mineral wealth would unfold a drama that would portray one of the most colorful chapters in building the nation. And while the setting became oriented to the advent of the dynamic and turbulent years, the Cinnabar Hills in Nuevo Almaden were yielding the treasure that would grow to proportions as one of the greatest quicksilver mines in the world during its years of operation.

In the early part of 1848, Alexander Forbes, accompanied by Thomas Bell, returned to Nuevo Almaden. Bell filled the position of confidential agent for the Barron, Forbes Company. In the later years when quicksilver became an important item on the industrial market, Thomas Bell would establish a lucrative business in San Francisco as an independent sales representative for the New Almaden quicksilver mines.

The return of Forbes was both gratifying and disillusioning. The daily operation was progressing favorably in spite of the inadequate facilities and inefficient methods. The crudely built retorts were producing a daily average of 100 to 150 pounds of quicksilver which was shipped to Tepic, Mexico. However, the picture became more distorted with complexity when informed by Walkinshaw that the existing conditions needed immediate attention. Forbes soon acknowledged that the Castillero claim was not sufficient in legal detail to show conclusively the absolute ownership of the mine.

The crux of the problem was the boundary agreement which had been mutually executed at the time of discovery. In 1842, Jose Reyes Berryessa and Justo Larios received land grands from Governor Alvarado. These grants were chiefly grazing land that extended into the hills. In the early establishment of the claim, Castillero covered the hill area in company with Berryessa and Larios and casually intimated the total area he desired. Starting from the point of discovery, Castillero spread his claim to four corners or roughly 3,000 varas. The hill area, considered worthless by ranch standards, became the outer boundary of the mine. Under conditions as they prevailed at the time and the casualness of the mutual agreement, there was little to indicate that a problem situation had been transacted. However, as time passed and the mining activity on the hill was revealing evidence which was creating great interest in certain parties, the attitude of the adjoining landowners began to take on a new character.

The situation was becoming more apparent that the unstaked claim of Castillero was an open challenge for the individuals who were questioning the validity of the claim. The strategic plan for developing a controversy into an open issue of contest was to establish the fact that the Larios and Berryessa land extended into the mining property. The widow Berryessa and Larios found the situation quite appropriate at this time to speculate on such a move. The Larios property claimed land into the area of the reduction works and the Berryessas laid claim into the mineral deposit.

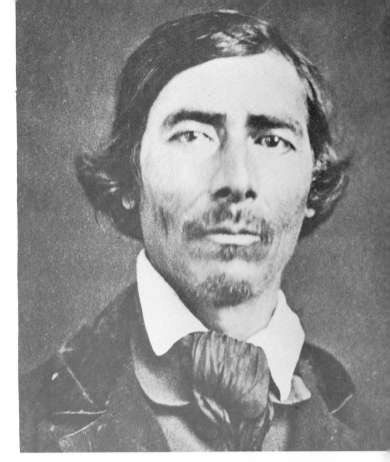

Charles Fossat, an opportunist, arrived on the scene and immediately recognized the advantages of acquiring the Larios property. In maneuvering tactics to gain this objective, he was assisted by a newcomer, of questionable scruples, named Henry Laurencel. The somewhat dejected James Forbes centered his attentions on the Berryessa land and began negotiations with the widow Berryessa to purchase the property. The first stages of the manipulative measures to be enacted in the procedure to gain control of the mine were agressively under way.

The future of the Barron, Forbes Company in exploiting the Cinnabar Hills became more hypothetical as the forces of opposition methodically entrenched themselves for a battle to the finish. Alexander Forbes now realized that they were facing a serious challenge and concluded that immediate action was necessary to counteract the tactics of claim jumpers, speculators and promoters of unscrupulous qualifications. At this opportune time, he was able to obtain the services of Chester S. Lyman, a professor of physics at Yale University. Mr. Lyman was a surveyor and when not engaged in the teaching field, spent much time in traveling and surveying assignments.

Chester Lyman recorded this account of his first visit:

"On February 7th made preparations to go out to the mine of New Almaden to make a survey of the lands granted them. Started late and reached the Hacienda at 7:30 p.m. Found Mr. Alexander Forbes and Mr. Walkinshaw in a small house recently built. Mr. Forbes is of the firm of Barron, Forbes and Co. of Tepic, Mexico, and British Consul there. He is an active, healthy, sociable man, apparently 55 though in reality 70. We were kindly received. The Mexican servant took our mules. Tea was soon ready and about 11 p.m. we retired, our beds being made on benches in one of the two rooms of the small house, the other room being the sleeping apartment of Messrs. Forbes and Walkinshaw."

During his stay of several weeks, Chester Lyman diligently covered the total area of the mining property. Any means of arriving at an accurate survey was purely hypothetical as the unstaked boundaries offered little upon which to establish a satisfactory conclusion. During his trips over the area, he attempted to establish the outlying boundaries as derived from a starting point which was the place of discovery. He was frequently in contact with the adjoining property owners of the Berryessa and Larios grants who attempted to enlighten the situation with their own solutions. Lyman concluded his work which resulted in a general disagreement with all parties concerned in challenging the validity of the Castillero claim and its extension into the adjoining properties.

As the pioneer operations of the Barron, Forbes Company became more promising, the value and stature of the mine became more obvious. The discovery and promotion of a quicksilver mine in California could not have happened at a more opportune time. The Gold Rush days had arrived as thousands of fortune seekers swarmed through the hills and sluiced the streams. The future development of the great mining era throughout the West would become a market for unlimited quantities of quicksilver. The success of the great gold and silver mines would be wholly dependent, during the early years, on the production of this valuable product. As the stamp mills crushed mountains of ore, quicksilver would serve its purpose to produce the bullion of wealth. This liquid metal served its purpose when applied to trays over which the finely crushed ore was washed. Having a natural affinity for uniting with gold or silver, the rich particles in the crushed ore were retrieved by the quicksilver, creating an amalgam. A firing process was applied to the amalgam which released the quicksilver by vaporization, leaving the desired product in its purest form.

With valid reasons to stimulate increased optimism for the mine and the future market that appeared inevitable, Forbes became more agressive in the immediate need to accelerate production. The ore was rich and plentiful but it would be necessary to develop facilities for extracting greater quantities. The small assemblage of retorts were serving the purpose but with limited success. In desperation to facilitate increased production of quicksilver, Forbes engaged Dr. Tobin from England to take charge of the reduction works.

During the early years, the miners were able to extract considerable quantities of ore that were obtainable with little development work. The ore bodies on the Hill were found entirely in Cinnabar and the deposits were discovered in large pocket formations. With minimum facilities and equipment, the Mexican miners were very efficient in their rather primitive type of operation. They possessed a natural aptitude and sagacity and were willing to risk any experience in the many precarious situations. Their ability to trace and discover ore bodies was a valuable asset in revealing the future potential of the Cinnabar Hills. Working with gopher-like precision, the Sonorians burrowed their way into the rocky slopes to create great open chambers. With no timbering to safeguard their method, large columns or stanchions were left to support the overhead weight. As one chamber was depleted of its ore, they tunneled their way with accuracy to locate another. Their method of operation was carried on by two groups of workers which were classified as Tanateros and Mineros.

The Tanateros were the carriers of ore which was the most severe in testing the stamina and physical endurance. They took a certain pride in their physical ability to bring out the ore over what generally was a tortuous course of travel. With his Tanate, supported by a strap passing over the shoulders and one around the head, the muscular Tanatero carried loads averaging 200 pounds. Day or night for a shift of ten hours, this human hoist transported the ore up the escalera through dimly lighted caverns and tunnels to deposit his load. The length of the laborious journey varied

but could involve the climbing of several escaleros of 10 to 20 feet in length and, under average conditions, 20 to 30 trips would be made during the shift.

The Mineros were the workers engaged in breaking down walls of rock with picks and crowbars. They were more skilled by their ability to identify and extract ore. Their work in the intense darkness of the cavernous chambers was carried on by candle light and small fires. The ore was extracted in upraises and downward inclines as they followed the cinnabar deposits. In certain formations, the ore was encased in hard rock or chlorite slate and it was necessary to use blasting methods.

As the mining operation continued into the year 1850, plans for improving inefficient methods and facilities became the major concern of the management. Under the procedure as practiced, the future days of ore production would be short-lived without immediate development work to promote long-term operation. The shallow deposits were becoming more scarce and less accessible. Deeper penetration into the ore bodies could not continue under the primitive methods and equipment which had served with a fair degree of success during the promotional days. Alexander Forbes, in anticipation of a promising future for the quicksilver market, recognized the imperative need for the reorganization of the operational structure.

In conjunction with the continued controversy, the Barron, Forbes Company had established a professional association with the San Francisco law firm of Halleck, Billings and Peachy. Captain Henry W. Halleck, a West Point graduate, was chiefly engaged in the lucrative field of California land titles which were much in dispute following the end of Mexican jurisdiction. His brief association with the Barron, Forbes situation and his legal understanding of Mexican land grants, created an inquisitive interest in the Nuevo Almaden quicksilver enterprise. He withdrew from his activities in San Francisco and accepted the request of Alexander Forbes to assume the managerial duties of the mine. As general manager and director of full-scale operation, Henry Halleck arrived in the Cinnabar Hills where he would be doubly engaged in legal and mining activity.

Accepting the situation of being the first general manager, Halleck approached the basic problems with plans designed for immediate change. The first project of concern was to improve the methods for extracting ore. The logical solution in accordance with mine development was the construction of a tunnel, deep into the promontory which would become known as Mine Hill. Located some 300 feet below the summit, this adit, which became the Main Tunnel, was started and in 1851 had extended a distance of 807 feet. It was constructed with an arched tunnel, 10 feet in height with heavy redwood timbers spaced two feet apart to support the walls. The width was adequate to accommodate a double track system for the coming and going of loaded and empty ore cars. At the entrance to the tunnel, a large area was filled and graded to serve as the depository for the ore. A long open shed was constructed which was called the "Planilla." Here the ore cars were unloaded and crews of laborers would break the ore to a specified size and segregate according to its value. In the grading process, the ore was placed in one of three groups referred to as gruessa, granza and tierra. The gruessa was the highest grade or purest quality; the granza was good ore but contained other rock or substance; the tierra was an inferior quality usually found in debris or refuse earth. Because of its fineness, the tierra was moistened and fashioned into brick forms and laid out for sun drying. Thousands of tierra bricks were stacked at the reduction works and during the winter season when the rain or road conditions curtailed the transporting of ore, they served the function of keeping the furnaces in operation. In the later development of the tunnel, access was available to many great ore bodies which were called "labores," and each became designated by a specific title. The most prominent that were located in conjunction with the tunnel were the San Pedro, San Rosalio, San Antonio, San Ricardo and San Pablo.

The grueling work of the Tanatero was replaced by the mechanical efficiency of a small steam hoist. Huge iron buckets would be hoisted from the dark, watery depths to deliver greater quantities of ore. The sturdy mule would contribute immeasurably as the heavy ore cars were drawn from the tunnel. The Mineros would be breaking the ore down with more rapidity and efficiency, as blasting replaced much of the hand work. Day and night, the shrill whistle of the steam hoist would alert workers, as the black powder explosions reverberated through the caverns with deafening roar. The darkness in the cavernous domain would be alleviated by the consumption of 70 pounds of candles every twenty-four hours.

The year 1850 was the beginning of organized operation in which primitive and nondescript methods were replaced by a systematic procedure. Ore production was accelerated beyond the capacity of the reduction works to maintain even terms. The retort method had become obsolete in the performance of increasing quicksilver production. Manager Halleck and Superintendent John Young designed and constructed an experimental furnace of brick and cement. The trial operation proved very successful in roasting the ore but was deficient in controlling the escape of noxious vapor. However, the first furnace was a definite step forward, and more construction was continued until 1854 when there were thirteen in continuous operation. Machine and blacksmith shops were built to serve the many needs and maintenance of equipment. On the Los Alamitos creek, a large Mexican water wheel was installed measuring 22 feet in diameter by 5 feet in width, which furnished power to operate some of the shop machinery. The successful operation of the wheel continued in service for a span of fifty years.

The first constructed furnaces were 6 feet wide, 10 feet high and 40 feet long and were arranged in a parallel formation 6 feet apart. Each furnace was divided into three parts consisting of the fire box, the ore chamber and the condensing unit. The fire box was located at the front and was separated from the ore chamber by a brick wall with openings for the passage of heat. The condensing unit was attached to the ore chamber by a brick wall with openings at the top. Each furnace would accommodate 15,000 pounds of ore and required 12 to 14 condensers through which the vapor is conducted, alternating above and below, until it reaches the end where it passes through a wooden cistern. The cistern is half full of water over which the vapor passes and cools on its way to wooden chimneys. Each condenser is fitted with a small pipe through which the vapor or quicksilver flows in streams along a narrow trough into pipes leading to iron vats. The quicksilver is deposited in the vats and from there is ladled to the scales where its weight is accurately measured and poured into iron flasks. The flasks were 18" in length, 8" in diameter and ¼" thick. Each flask contained 75 pounds of quicksilver and was sealed tightly with a screw-in plug. During the early years of operation, the flasks were imported from England.

The furnaces received the ore from the top and were carefully arranged by a worker in such a way as to get maximum heating. The top of the furnace was well covered with mortar and dry ashes to prevent the leakage of vapor. In spite of the precautionary measures, there was always a certain percentage of vapor loss in the form of arsenic with sulphate of mercury. The workers at the furnaces were confronted with a hazardous occupation from the standpoint of health impairment. The fumes from the mercurial ore caused salivation which had a deleterious effect upon the physical well-being of the workers. It was customary under these precarious conditions to rotate the crews by which they were engaged at the furnaces one week out of every four. The shifts ranged from 10 to 12 hours for six days at wages of $2 to $2.50. The average time required to roast and extract the quicksilver from solid ore was about 56 hours. When filled with tierra adobes, the time was generally 50 hours.

Under the capable direction of Henry Halleck, the revised pattern of procedure gathered considerable momentum in development of the mine and production of quicksilver. Profits were reduced by the continuous investment in property and development work. The ore to the furnaces was yielding an average of 36% quicksilver and the operation was showing conclusively that the mines of Nuevo Almaden had a great potential.

From Mexico, South America and the mounting activity of the gold fields, came increased demands for quicksilver. Teamsters with their wagons loaded with flasks left the Hacienda at regular intervals for the little shipping port at Alviso slough. Here, the cargo was stored in a warehouse along with hides and tallow for delivery to San Francisco. During the mission days, the place was known as El Embarcadero de Santa Clara Asis. When Ignacio Alviso, majordomo at the mission settled here in 1840, the settlement took his name as the title of the locality.

During the years 1850 through 1863, Alviso played an important role as a terminal point for freight and passenger service by water to the Santa Clara Valley. The marshy area was surveyed by Charles Lyman in 1850, and a townsite was staked out with lots listed at $600. There was considerable optimism that the future years would find the settlement thriving with business and population. Peter Burnett, the first governor, had a house built here, but when developments failed to materialize, he had the house dismantled and had it rebuilt in San Jose.

The first steamer to travel between Alviso and San Francisco was a converted scow, the Sacramento. Other boats that became a part of the service were the Firefly, New Star, Boston and Jenny Lind. Passengers were charged $40 for a one-way ticket. Stage service carried the passengers from the boat on a course through Santa Clara and over the tree-lined Alameda to San Jose for a $5 fare.

The anticipated prestige for Alviso never came as the promising outlook diminished abruptly in January, 1864 with the completion and immediate service of the San Francisco and San Jose Railroad.

On the Hill, in Deep Gulch, the settlement continued to increase in size as more dwellings were built to accommodate the steady arrival of new workers. While the population was predominately Mexican, other nationalities were attracted by the publicized activities in the Cinnabar Hills. The general environment lacked considerably in any degree of favorable living standards and with the crude conditions that prevailed, there was little incentive for improvement. The quickly-built, unpainted houses and the raggedly dressed people, children in scanty attire, with a fair representation of dogs, pigs, poultry and donkeys, composed the typical picture. These early inhabitants adjusted to an environment that was uncontrolled and at times violent, as law and order was yet to establish its authority in this new settlement.

During the years of the Barron, Forbes operation, until the change of ownership in 1863, the population was without benefit of medical service or any facilities for health or personal welfare. Many died from common diseases, accidents and other contributing causes. The company showed little concern for the needs of the workers or the improvement of living conditions. The only medical service available was in San Jose, a distance of 12 miles. Such service was difficult to obtain because of the distance and the lack of communication. Also the inability of many patients with meager finances to meet the medical cost, added to the adversity of the situation. Doctors were reluctant to make the round-trip which involved three to four hours from their practice for little realization of remuneration.

Dr. W. S. Thorne, who in later years would become a resident doctor at the mines, gives this account of his visit to Nuevo Almaden in 1858:

"Upon my first visit to Nuevo Almaden in 1858, the general environment and its population was not too favorable. The general conditions in the settlement were somewhat unrestrained, turbulent and with little concern for law and order. Many renegades, murderers and thieves were frequent visitors to the locality. However, the standard of living was comparable to other mining camps throughout the West. During the early years, most of the workers were peons from Mexico, and renegades from justice. Almaden was a publicized rendevous for some of the worst element in the state."

The Barron, Forbes Company were fortunate during their period of mining to have had an abundance of rich ore within shallow workings. The prolific distribution of ore accounted for little underground development and the company investment in the enterprise was related chiefly to dwellings, workshops and furnace yard facilities. One of the main and last projects of any importance was the construction of a tunnel into the hill which would drain off water at the 600 foot level. This work started in 1857 and was known as the New Tunnel. Sherman Day, a civil engineer, was the superintendent and surveyor for the company who had arrived in Nuevo Almaden in 1856. He supervised the construction of the tunnel until the day in 1858 when all work stopped because of the injuction placed upon the property. In later years following the lifting of the injunction and the resumption of mining, the project would be continued under a new title called the Day Tunnel.

With the termination of activity, most of the employees were off the payroll. Many of the Mexicans remained in the settlement as the company did not anticipate a lengthy controversy. Others, who were capable of moving to new fields gathered their belongings and departed.

Sherman Day left to become engaged in surveying work in the area of Folsom. An excerpt from a letter to his father, Rev. Jeremiah Day, President of Yale University, indicates briefly, the prevailing situation in his place of employment:

"On Wednesday, November 16, we packed our trunks and left our beautiful home at New Almaden, to wander like gypsies, in search of other employment. The U. S. District Court issued an injunction against the abstraction of any more ore from the New Almaden mine until some more definite ascertainment of the title should be had. The reasons for this action you will find in a pamphlet I sent by mail to Mr. Lyman. One side of the Court wants to appoint a person to carry on the business on account of unknown ownership. The mines, therefore, are likely to be suspended for many months, perhaps years, and my occupation's gone in that locality."

As California made the transition from Mexican to American law, many land grants to settlers by the Mexican government, became an issue of much controversy. The situation developed into proportions requiring the appointment of a Board of Land Commissioners in March, 1851, by President Fillmore. It became mandatory that any land owner, by Mexican grant, file claim within two years or lose his rights to the property. The New Almaden Mining Company presented their claim in 1852 to establish the legality of the Castillero title. It appeared that the documents presented were insufficient and contained contradictory dates which gave evidence of fraudulent and false statements. Under these conditions, in spite of the favorable reaction by the Land Commissioners, a suit began which would carry on through the District Courts to the U. S. Supreme Court over a period of twelve years. The unstaked claim of Castillero and the adjoining properties of Jose Reyes Berryessa and Justo Larios, became a complicated issue and the lengthy procedure attracted national attention.

A federal injunction was enforced in November, 1858, against the New Almaden Mining Company. Moving into the legal conflict on behalf of the company, was the law firm of Halleck, Peachy and Billings. Judges Hoffman and McAllister, presided in the United States District Court and in conclusion of the case, substantiated the Castillero claim in certain parts and rejected others. The case was appealed to the Supreme Court and the contest continued under the title of "The United States vs. Castillero."

Following the appeal in January, 1861, to the highest court, the New Almaden mines continued to operate in spite of the injunction. During the year 1862, the Supreme Court ruled against the Castillero claimants by a margin of 4 to 3 and later in May, 1863, President Lincoln, signed an order to evict the Barron, Forbes Company in a proclamation that the land was public domain and subject to government control. Leonard Swett, was sent to New Almaden as a government agent to take over the property. He presented his order to C. W. Rand, a U. S. Marshal, to enforce the action. Swett arrived at the mines accompanied by Samuel Butterworth where they were confronted by a riotous and militant assemblage of armed miners, who readily made known their intentions to stay on the property. Recognizing the hopeless situation of gaining admittance to the property, Swett requested government troops be assigned to the locality should stronger measures be necessary to carry out his assignment. The rather explosive situation subsided when the newspapers incited public indignation by their caustic and critical comments of the government initiative to sieze mining properties. President Lincoln was besieged by letters and telegrams from a wide range of influential individuals to withdraw the order. Lincoln made immediate acknowledgement and the situation at New Almaden was relieved.

The litigation proceedings came to a close in April, 1864. The contestants under the Fossat grant, which was the Quicksilver Mining Company of New York and Pennsylvania, received the favorable verdict. However, the victory was not sufficient to assume outright possession. The company in order to gain full legal ownership was forced to enter into a financial transaction to purchase the existing 8,580 acres and also the cost of the property developments expended by the Barron, Forbes Company. The deal was mutually concluded at a price of $1,750,000. The arbitrary line established by the Supreme Court, placed the mines on the Los Capitancillos and the reduction works on the Rancho San Vincente. Both properties were combined and the entire holdings became the Quicksilver Mining Company.

Hubert Howe Bancroft in his History of California, Volume VI, makes this reference to the case:

"Of famous cases, the claim of Andres Castillero for the New Almaden Mines was probably the most important and complicated. In magnitude of interest involved and bulk of record, this case before the District Court was deemed second to none decided previously by any tribunal. The transcript of record filled 3,584 printed pages; 125 witnesses were examined, 18 of them prominent men from Mexico; lawyers like Reverly Johnson, Judah Benjamin, Hall McAllister and Edmond Randolph on one side or the other gave utterances to from 100 to 400 pages each of legal lore, eloquence, wit and sarcasm. Dozens of special pamphlets on the subject were published besides the regular Court records; and outside of the main struggle between claimants and the United States, there were always complicated litigations between the quarreling claimants. The great battle had to be fought again before the Supreme Court, where, by an unjust decision, the mining claim was finally rejected; and another struggle in behalf of a survey that should locate the mine on private lands controlled by the company, the latter was forced to yield and part with its property at a nominal price of $1,750,000."

The Barron, Forbes Company having dissolved its interests in Nuevo Almaden, returned to Mexico. Their short span of years in promoting the first legal mining claim in California, had been highly successful. With a minimum investment in development work, they had produced quicksilver valued at $15,000,000. They pioneered a mining venture that would continue to flourish for a half century and established a settlement still known today, as New Almaden.

CHAPTER 3

SAMUEL
BUTTERWORTH

The Quicksilver Mining Company

Moves Into New Almaden

———

With the culmination of litigation and a general settlement of the controversial issues, the New Almaden mines faced a new beginning. During the years of the injunction which legally closed down operations, conditions, in general, had deteriorated to a low standard. With the departure of the Barron, Forbes Company and the immediate possession of the Quicksilver Mining Company, a complete reorganization plan became the primary objective in the resumption of operations.

The individual, who played the major role in the financial negotiations that acquired ownership for the new company, was Samuel F. Butterworth. He was president of the newly formed company, with offices in New York, and incorporated under the laws of Pennsylvania. During the time of litigation, the New Almaden mines received sufficient publicity to interest speculators in such an enterprise. In the Spring of 1863, Samuel Butterworth made a special trip to the Cinnabar Hills for a critical observation and to survey the mining potential as a major investment. Should the outcome of the litigation, place the property in a situation whereby, it could be purchased, Butterworth would have the advantage by his own personal evaluation and a solid basis for any financial negotiations. It is quite possible that an item in an article, "The Quicksilver Mines of New Almaden," written by William V. Wells and which was published in Harper's New Monthly in 1863, could have caught the eye of Mr. Butterworth. An excerpt from the article gave this appraisal (1857):

SAMUEL F. BUTTERWORTH

The first president of the Quicksilver Mining Company was Samuel F. Butterworth who came to New Almaden in 1864 and assumed management of the mines until July, 1870. Following his arrival, he established the first store in the Hacienda and one on the Hill in Englishtown both of which would continue to serve almost completely the needs of the settlements for the duration of the mining activity.

———

"The ore of New Almaden is solely sulpherets of mercury, the rarest known and exceeds in richness that of any other on record. The average yield is 35%. No native or virgin quicksilver has yet been found. The general outlook of the New Almaden mines at this time would indicate that the mine is inexhaustible."

Sam Butterworth spent some time at New Almaden and conducted an analytical study of every aspect of the settlement, mines and equipment. He was convinced that the red earth contained a great potential and that the outlook for long term operation showed great promise. He, also, gained first-hand information of the basic details which had developed the controversy and was, therefore,

THE HACIENDA IN 1867

This is one of the earliest photos of the Reduction Works and the settlement of the Hacienda in the background. The thoroughfare starts with the General Store and continues with the miner's cottages, many of which are still in use today. The trees fronting the houses are in early growth.

(Watkins Photo)

well informed as to the extent of the mineral boundaries. With a complete record of the necessary details, Butterworth returned to New York and prepared his report for the Board of Directors.

The enthusiasm of Sam Butterworth and the excellent report convinced the Board that they must exert every means of strategy to negotiate the purchase of the property. Butterworth worked diligently and made a strong impression with the group who was contesting the Castillero claim. He made known the desire of his company to purchase the property should the situation prevail. The final decision of the Supreme Court created this possibility, and the involved parties relinquished their holdings by sale to the Quicksilver Mining Company. It was the decision of company directors that Sam Butterworth move to New Almaden as General Manager in complete charge of the total operations. He resigned his office as president and accepted the new position at the annual salary of $25,000.

Samuel F. Butterworth was born in Newburgh, New York. After graduation from Union College,

he entered the practice of law. During the early years as a successful lawyer, he was attracted to the state of Mississippi for what appeared to be a lucrative field. He established associations which proved advantageous in his new environment and his work progressed with favorable success. However, the day arrived when his Yankee presence was scrutinized as unbecoming to the southern state and some of the "toughs" and duelists decided that he should depart. A reputable duelist was chosen to involve Butterworth in a situation of conflict and through challenge and insult, he was obliged to face his adversary or leave the state.

On a chosen day, Butterworth arrived at the scene where a large crowd had gathered to witness a traditional means of settling differences. Both contestants were presented with a revolver for one hand and a knife for the other. The site was measured at ten paces and at a given signal they were to fire from a marked base line. On signal, both men began firing but with little success in inflicting damage; whereupon, Butterworth's antagonist discarded his revolver and rushed madly forward to complete the affair with his knife. At this stage, the incident was brought to a close as the crowd, not relishing a knife fight, surged in and stopped the encounter. Butterworth, inexperienced and unaccustomed to such practice, considered his experience most fortunate to have escaped unscathed. Within a short time following this event, Butterworth packed his belongings and returned to New York.

The practice of law only partially satisfied his desires and abilities and very soon he became attracted to politics. He became an active worker in the Democratic party and his vehement nature, expressed in a cool and reflective manner, established his capabilities as a natural leader. He was inflexible in his business transactions where he was a bitter enemy to his adversaries and a faithful friend to his associates. During the Van Buren administration, he was appointed U. S. District Attorney in Mississippi and before the close of his term, was commissioned Justice of the Supreme Court in that state, an honor which he declined.

During the time following his return to New York, the much publicized New Almaden litigation had been active and as previously stated, Sam Butterworth was selected to venture into a new horizon to assume the responsibility of building up a deteriorated mine. The selection was both timely and wise because of his training, personal qualifications and background of experience. He was a proven candidate for rehabilitating and reorganizing the chaotic situation in New Almaden. His keen understanding and successful experience in financial transactions gave the company the desired faith in the capabilities of their general manager.

In July, 1864, Samuel Butterworth arrived in New Almaden with carefully laid plans for an immediate transformation of existing conditions. Within a short space of time, his agressive manner was applying forceful measures for increased efficiency of operation and improvement in the general welfare of the mining community. Systematic routine and the establishment of complete authority became the keynote in the operations of the dynamic Mr. Butterworth. Rules and regulations were imposed and every employee became oriented to the specific details of living and working under the new management. The standards of operation were raised to facilitate greater production and only the qualified and competent workers were retained on the payroll. By the close of 1864, a remarkable advance had been made with the production of 42,489 flasks of quicksilver which were shipped to the open markets in China, South America, Mexico and the mining mills of Nevada and California. At the market price of $45.90 a flask, the Butterworth management had grossed for his company $1,950,345. The largest number of employees to ever work at New Almaden were engaged at this time, and the total wages distributed amounted to $26,097.

The initial performance of Sam Butterworth clearly indicated his abilities and still greater progress was achieved during the year 1865 when production reached the all-time high of 47,149 flasks which, at the same rate, amounted to $2,166,304. The year 1866 was also commendable in spite of the very noticeable decrease in production, with 35,150 flasks. The price had increased to $53.15 which grossed $1,881,759. The first three years of operation by the Quicksilver Mining Company attained the peak of production and income which was never equalled during the life-time operation of the mine.

As the Cinnabar Hills hummed with activity under the forceful measures of a rigid routine, Butterworth was, also, concerned with the improvement of community welfare. The isolated settlement was at a disadvantage in respect to the needs of the inhabitants, and the area was an open market for the sharpers, speculators, quacks and peddlers. The meager earnings of the workers contributed generously to the livelihood of the unscrupulous solicitors and their unrestrained practice. The old adobe warehouse constructed by Barron, Forbes was reconverted into a merchandise store. On the Hill, a sturdy brick structure was erected to serve the miners. With an assorted range of merchandise offered at reasonable prices, the inhabitants were able to select and purchase at their own discretion. Following the opening of business, the stores were leased to C. J. Brenham and Thomas Derby.

The left-over remnants of the Barron, Forbes days were diminishing rapidly as the remedial methods of Butterworth consistently reshaped the environment for complete dedication in serving the company. The company proclaimed itself a private industrial institution and the mandatory policies, put into force, allowed no exceptions to any rule. The mining area was posted as private property and the boundaries were fenced to control the traffic of visitors, peddlers or trespassers. Unlike the open domain of most mining camps, the mine settlement of New Almaden assumed all the appearances of a feudal estate. A toll gate was established at the entrance to the property and outsiders were admitted by an attendant only upon presentation of an official approval. This practice was rigidly enforced to exclude certain elements

THE TRAMWAY

This rail system for transporting ore from the Hill to the furnace yard was constructed under the direction of Samuel Butterworth in 1864. This added feature decreased cost and greatly increased ore production to the furnaces. From the base of the incline, the ore continued by rail, drawn by mules to the furnace yard. The loaded cars coming down pulled the empties back to the top. This was the scene in 1890.

which had previously created undesirable situations by their presence. Except for a certain number of nocturnal visitors by remote trails, the toll gate entry controlled this situation for the many years that followed, during the operation of the mines.

One of the most important innovations, initiated by Butterworth which contributed immeasurably to increased ore production, was the installation of a tramway system on the slope of the hill which descended from the mine area to the furnaces below. The ore cars were in continuous operation day and night and many loads of ore were delivered with minimum time and labor. The heavy, slow-moving wagons, that had faithfully served during the early years in transporting the ore over a two mile tortuous journey, became obsolete. The tram system proved highly efficient and its simple manner of operation, requiring little maintenance, continued to serve the furnaces for the remaining years of company operation.

During the early years, the population of youth had been totally neglected as the remote settlement offered little to their welfare for mind or body. In 1864, the number of children was sufficient to warrant consideration, and a school district was established within the boundaries of the mining property. As most of the population was concentrated in Spanishtown, a one-room building was constructed and the three R's became a challenge to the undisciplined youth.

By 1866, Sam Butterworth had initiated the basic projects of his reorganization program. There were still problems and improvement in operations would be a major objective during the life-time of the mine. With the pattern definitely set and the general conditions having leveled off to a fair standard of achievement, Butterworth opened a company office at Front and Jackson Streets in San Francisco. His superintendent, who had been well indoctrinated in the policies which had been enforced, was put in complete charge but with strict adherence to the orders of the general manager. Between the Casa Grande residence in Hacienda and the financial world of San Francisco, Butterworth maintained a consistent schedule of personal contact and supervision. He assumed, more and more, a new role of giving orders rather than enforcing them. For the remaining years of his office as general manager, he would rely almost entirely upon the efficiency of his superintendents.

During the relatively short tenure of Butterworth, seven individuals served as superintendent for a short duration of time. James Eldridge was the first, followed by Sherman Day, N. S. Arnot, James Nowland, E. J. Mayo, C. E. Lightner and C. E. Hawley. One of the most capable men was Sherman Day who returned to the scene where he was engaged during 1856-58. He, like many others, was forced to leave when the injunction was applied to close operations. Day resumed his work in April, 1864, but by November submitted his resignation. His short stay and reasons for leaving was his inability to work and communicate with the Mexican workers on any acceptable basis. Day was instrumental in developing the tunnel which was later named in his honor.

Probably, the most conspicuous individual to serve in this capacity was James Nowland. He was well indoctrinated with the Butterworth formula for strong discipline which he tried to enforce in a rough, boastful and swaggering manner. He was an early prototype of the Western gunslinger, and

THE MAIN TUNNEL

During the early development days of the New Almaden mines, most of the ore was brought from the depths of Mine Hill through this tunnel which opened on Deep Gulch. In this early day picture, miners are emerging with a full load of cinnabar for the sorting shed. At the extreme right is Sherman Day, mine superintendent during the 1860's.

each day he was garbed with two holstered revolvers which he wore to better emphasize his authority. Jim Nowland, while fairly efficient in mine operation, cared little for popularity from the workers. His method was to handle all situations according to his own solution and that which would coincide with the policies of the management. Whether he liked the superficial stature he derived from his attire which gave him security or because of his harsh, impersonal and forceful nature, was never known. Off the job, Jim Nowland assumed a debonair characterization and was a popular associate of the more genteel society. His presence was always conspicuous at the social affairs in the Hacienda.

Jim Nowland survived many problems and, during his days on the job, he kept vigilant watch over the mining domain and, also, by correspondence informed his superior, Mr. Butterworth, of the routine matters in New Almaden. The following letter gives some indication of Nowland's manner and the methods by which he attempted to establish complete authority:

S. F. Butterworth, Esq   January 30, 1866

Dear Sir,

I must call your attention to the fact that there are yet on the Hill, some of the leaders in the strike of a year ago. The villain Mariano Cayala has lately returned. Jose Antonio has been here on a visit. I am satisfied they are brewing mischief.

I am obliged to go to San Jose on Thursday to answer the charge of assault and battery on a Chinaman whom I ejected from the premises on the complaint of (I can't make out his name). I have dismissed Patterson for not attending to his duty. I wish to eject about ten men with their horses, etc., from the premises of the Company. I propose to take the upper hand myself at once and keep it. I must have some aid to enable me to do it. I think Chief Burke (sheriff in San Jose) would let you have two good men for two months. I can accomplish all I want and keep the employees in subjection.

21

The difficulty that I have to contend against is that concessions have been made before. I don't propose to concede anything but unconditional surrender, unless by instruction from you.

Yours respectfully
James A. Nowland, Supt.

In April, 1867, Jim Nowland realized that his position was a daily challenge which brought little reward, so he resigned and left the scene.

The last superintendent to be engaged by Butterworth was C. E. Hawley who was well informed in the details of mine operation. He handled the daily routine with intelligence and efficiency and concentrated his efforts to the best interests of the company. An excerpt, from one of his mining reports written in 1865, gives an interesting account of his observations on the Hill.

"At the mines is a village of considerable extent and over the whole ridge are scattered the houses of the mining population. Upon the estate are over seven hundred buildings, most of which are dwellings. At the mine is a population of about 1800 souls and at the Hacienda are about 600 more. The total number of men at work for the company is about 1,000.

"From the mine ridge to the northwest, the view is unbroken to the mountains of Marin county beyond San Francisco. The City is hidden by projecting hills, but Angel Island and the waters of the Bay are distinctly visible. Occasionally, the smoky train of the ferryboats can be seen floating away over the waters and fleecy clouds streaming through the valley, show where the iron horse is rushing over the land. Though sixty miles away in an air line, on still days the firing of the heavy guns in the harbor fortifications can be heard, reduced by distance, to a strange, low humming sound."

One of the prominent personalities who personified the true spirit of dedication to man and employer, with over forty years of service, was Charlie O'Brion. Young and inexperienced, he arrived at New Almaden in 1865 and started as a laborer. Charles F. O'Brion always made it a point, when referring to his name, that it was spelled with an o and not with an e. This energetic and amiable Irishman adjusted well to the raw, crude environment of a mining camp which was predominately Mexican. As the years passed, O'Brion settled himself more permanently to the life in a quicksilver mine, and he served the company as a night watchman in patrolling special areas, weigher of ore loads, foreman of the laboring

crews at the planilla and general utility man, wherever the situation called for immediate attention.

Charlie O'Brion grew with New Almaden, to become the pioneer employee with the distinctive title of surface foreman. He was the overseer of all the company property and his work involved a wide range of handy-man activities and jack-of-all-trades. Once a month he called on the company houses to collect the rent, which was a horseback tour about the hills. He supervised the wagon roads and kept them in repair, supervised the cleaning, recorded and shipped the ore to the furnaces, attended to the sanitary conditions of the settlements and was in charge of the company buildings and community recreational affairs.

In 1869, a much anticipated solution to the health and welfare for the settlements of New Almaden became a reality by the arrival of a young medic, Dr. W. S. Thorne, to become a visiting doctor for the mining company. The increased population and the interruption of operational routine because of health and accident problems accounted for the immediate action of Sam Butterworth in initiating a health service.

Dr. Thorne knew mining camps, the people and their problems. Following his arrival from England in 1857, he worked as a laborer in the camps of the Mother Lode and Nevada. In 1858, he visited New Almaden which was a raw settlement of 1600 inhabitants, existing under sub-standard conditions, socially and economically. The place was infested with an undesirable element and general health conditions were maintained on primitive and home-made remedies.

Upon his return ten years later, he found considerable improvement had taken place and in comparison to other mining camps of his acquaintance, he felt that New Almaden was, by general appraisal, superior in its social, economic and moral standards. The availability of a doctor, with supplementary facilities to accommodate the needs of the populace, was great news for a rather under-privileged community. At the store on the Hill, an apothecary shop was a feature, with a quality of drugs and prescriptions comparable to other similar establishments on the coast. The products and services were dispensed to the customer upon recommendation of the doctor at cost to all employees.

During the latter years of Sam Butterworth's management, the general direction of affairs was conducted from his San Francisco office. The outcome of the directives depended entirely upon

THE DAY TUNNEL AND PLANILLA

This operation was first called the New Tunnel. After resumption of work following the injunction, it became identified as the Sherman Day Tunnel in honor of the mine superintendent who directed the development of the project. This tunnel extended into Mine Hill for a distance of 1887 feet.

Courtesy of New Almaden Historical Society

the sincerity and ability of the superintendents and their subordinates to fulfill specific obligations. The production and income showed a marked annual decrease. J. Ross Browne in his report, "Resources of the States and Territories West of the Rocky Mountains," presented in 1868, made this reference concerning the situation at the New Almaden mines:

"The present conditions of the principal mine is poor, both in quality and quantity of ore, its future is uncertain and any conjecture would be valueless."

In all probability, Sam Butterworth was alerted by this statement which accounted for his diminishing interest after such a vigorous beginning. The background experiences of his career always seemed to indicate a great enthusiasm for the promotional aspects rather than long term operations. The spontaneous inclination to seek greener pastures and once again meet a new challenge was the beginning of the end for Sam Butterworth and the quicksilver business. He had fulfilled his primary objectives in promoting a successful mining enterprise for the Quicksilver Mining Company. The Hill still had a future, for they had only scratched the surface.

A letter of resignation was prepared and mailed to the directors in New York in April, 1870. He took the exception in a letter of this type to recommend his successor in the person of James B. Randol, his nephew and secretary of the company for seven years. Within a month, his resignation was acknowledged with this reply:

The Quicksilver Mining Co.
21 Nassau St., New York, N.Y.
May 9, 1870

S. F. Butterworth, Esq.
Manager,
Quicksilver Mining Co.
Jackson and Front Sts.
San Francisco, Cal.

Dear Sir:

Your letter of the 27th of April came duly to hand. The Board of Directors have concluded to accept your withdrawal from the management of the company from June 1st next, regretting that the product of the mine will not admit of retaining your valuable and watchful services.

Very respectfully yours,
John K. Pruyn, President

Following the acceptance of his resignation, Sam Butterworth closed his San Francisco office and spent the remaining days on the job at the Casa Grande residence in the Hacienda. James B. Randol arrived in New Almaden as the newly appointed manager at the annual salary of $12,000. He spent about a month in becoming indoctrinated to the life in a mining camp, the details, problems and over-all picture of conditions, after seven years of Butterworth's management.

In July, Sam Butterworth vacated his residence at the Casa Grande and James B. Randol became the new official tenant. Butterworth left the Cinnabar Hills to become president of a new development known as the North Bloomfield Gravel Mining Company from which he retired within a year.

Upon retirement from his many business activities, he was entertained with a touch of distinction by William Ralston, San Francisco banker and financial promoter, at his lavish estate in Belmont. The California years had been rewarding, not only financially but in other channels of recognition. One of his distinguished appointments was by Governor Haight to the Board of Regents of the newly-established University of California. He, also, served as chairman of the Executive Committee during the infancy of the University. In another phase of public service, he served as Commissioner of the Golden Gate Park.

The career of Samuel F. Butterworth came to a final closing in May, 1875. His active years in law, mining, real estate, financial transactions and politics, had culminated with achievement and success. To the very end his restless spirit was soothed by challenge and activity. His bluntness, strength and courage, expressed by a personality of force and strong will, stimulated his endeavour throughout his career to expend energy beyond his competitors. He left no monument or edifice to commemorate his name. The final tabulation of his earthly achievement was an estate valued at $7,000,000.

CHAPTER 4

JAMES B. RANDOL

Bonanza Years Establish

a Unique Mining Settlement

In July, 1870, James Randol, left his position as secretary of the Quicksilver Mining Company at the New York office and arrived in New Almaden as the newly appointed general manager. He was thirty four years of age with no actual experience in industrial management and completely lacking in the technical knowledge of mining operation. The appointment of an inexperienced person, to the position of directing and promoting the future of quicksilver production, was an optimistic venture of speculation on behalf of the company directors. While the general conditions of the settlements and quicksilver production had shown degrees of progress under the Butterworth management, there was, still, much to be done to insure any future for the operation.

The last several years, under Butterworth, had witnessed a certain depreciation in efficiency of production and a marked decrease in income. Operations had been expensive and the company faced a financial situation of indebtedness that threatened survival. The most immediate problem at hand was the liquidation of an interest bearing note of $1,600,000 originating from the initial purchase of the property. The stock was unassessable and the imperative need for funds required urgent attention. Randol approached this first major problem by soliciting financial subscription which attracted a favorable response of $200,000. This emergency contribution provided a working fund which subsidized the continuance of development work and made possible a successful enterprise that would flourish for the duration of company operation.

James B. Randol was born in Newburgh, New York, in January, 1836. His formal education was brief with emphasis in business training which became his vocational field. He was an imaginative

JAMES BUTTERWORTH RANDOL

After seven years of service as secretary for the mining company in the New York office, James Randol was promoted to general manager in July, 1870. A young man of 34 years and with no practical experience in mine operation, he came West to New Almaden where he successfully promoted the mining activity through its most prominent years. In conjunction with his official duties he contributed greatly to the improvement of community life and the welfare of the employees. He resigned his office in March, 1892.

---

and creative individual that functioned in business transactions with logic and practical solutions. His seven years as company secretary had enlightened a receptive mind to the prevalent problems in the Cinnabar Hills and from the standpoint of good business discretion, he was aware of the changes needed to establish a more progressive economy.

Upon establishing residence in the Casa Grande and assuming the challenging obligations of general manager, Randol proceeded immediately with pre-conceived plans that would effect every phase of living and working. The first concern was to advance the efficiency of furnace operations and to increase quicksilver production on a more economical budget. The furnaces, at this time which had served with a fair degree of success, were crude and inefficient by acceptable standards. They were called the intermittent type, which were capable of roasting only coarse ore and adobe bricks. The process was interrupted by in-

These three prominent Almadeners are left to right: George Lighthall, School Principal and Justice of the Peace, J. Wilkenson, Company Surveyor and Francis Meyers, builder of the Casa Grande.

tervals of inactivity due to the cooling off and extraction of roasted ore. These delays caused considerable loss of time and, also, a fair percentage of quicksilver due to the inefficient condensor system.

Randol was informed of the furnaces at the Idria mines in Austria which were designed to carry on a continuous operation of roasting and extracting with no intervals of inactivity. He engaged H. J. Huttner, a mechanical engineer and a brick mason, Robert Scott, to design and build a comparable device. It was during 1874 that the first experimental furnace was completed ānd put into operation. The results were highly rewarding in efficiency of operation, greater production, and with minimum loss of quicksilver. The furnace functioned day and night with a continuous feeding of ore at the top and when roasted, was extracted at the bottom. Less fuel was required because of the consistent maintenance of temperature, with no cooling-off period. Construction of the new in-

novation continued and eventually, five furnaces were operating day and night and roasting an average of 145 tons of ore every twenty-four hours. The ingenuity of the designer and builder had produced one of the best facilities ever devised for the reduction of quicksilver and they continued in operation until the end of the mining activity. This very vital improvement came at the most opportune time for without increased production, the continuance and future activity in New Almaden was doubtful.

Faced with the problem of diminishing capital and increased costs of operation, Randol realized the necessity for better methods and facilities in order to develop the potential wealth within the contours of Mine Hill. The extraction of ore, on the 800 foot level, required considerable handling which depended chiefly on manual labor. On June 10, 1871, Randol initiated the construction of a new shaft which would serve to deepen exploration and also, a better means for hoisting. Lack of foresight resulted in a shaft which was smaller than prevailing standards. Within an area of 4 feet by 9 feet, the space was only adequate to contain a single hoist which through the years of operation restricted the production of ore. The maximum output of ore from this shaft averaged about 300 tons per day and with accelerated developmental work in full swing it was not possible to stop work to enlarge by alterations. This new entry into Mine Hill was known as the Randol shaft and for twenty years maintained a prestige as the most important shaft. Relief was brought about for the Randol shaft by a new entry named the Santa Isabel, which was commenced in 1877. This structure was large in size with accommodations for three hoisting compartments. Large pumps were installed and gave much needed relief for water drainage from the Randol workings.

Conditions soon became apparent with everyone, familiar or associated with the mines, that the new management was administering effective measures for promoting higher standards and developing the mineral potential for long term operation. This progressive leadership gradually created a loosely governed operation into a smooth running, systematic routine that carefully controlled extravagances and raised the standards of the working conditions. Incompetent and undesirable elements were eliminated and Randol, greatly impressed by the stability and working techniques of the Cornish miners, encouraged an influx of these people from Cornwall and the mining camps of the Mother Lode.

Under the critical and precise supervision of James Randol, the settlements became adjusted to

QUICKSILVER MINING COMPANY OFFICIALS

This picture, taken in 1889, shows the Top Brass who conducted the official duties of the mining company. (Front Row): L to R — John Martin, Shift Boss; Dr. J. Underwood Hall, Physician; Sidney Jennings, Surveyor. (Back Row): L to R — Charles O'Brion, Planilla Foreman; James Harrower, Chief Engineer; James Varoe, Mine Foreman; Captain James Harry, Senior Mine Foreman.

the authoritarian attitude of the company. The general decor of the environment was elevated beyond the average standards of most mining camps. The dwellings scattered about the rough terrain, enclosed with picket fences, glistened with white paint. The humanitarian motives of the new manager was obviously conveyed to the inhabitants by the interest in creating congenial accommodations that gave a feeling of permanency to the people in an environment that by any comparison, was unique in the life of a miner. The company maintained the homes with necessary repairs. Plants and shrubs were furnished from the Casa Grande estate.

The new arrivals entrenched themselves in their environment and for the majority, the Cinnabar Hills became their home for the duration of the company management. The idealistic viewpoints of Randol, to create a congenial and compatible way of life that contained the amenities for pleasant living, was apparent during his years of service. His interest in the welfare of the company settlements, developed a certain dedication that through the years was expressed in many practical ways.

One of the most imperative needs, which became the personal concern of James Randol, was the inadequate service for health and hygiene. The young growing population and the increase in childbirth had created demands beyond the partial offerings of a visiting doctor. The previous years had revealed little evidence or concern of any established facilities for the wide diversity of health and accident cases. The first medical service was administered by Dr. W. S. Thorne who came to the mines on specified visits each week. While the partial service was beneficial to a limited extent, it was not sufficient to handle the increasing demands of the several settlements and the need for a full time resident doctor became an urgent necessity.

In 1871, Randol, after conferring with many of the employees, conceived and organized a health welfare plan that would provide medical service to all the employees and their families. The plan was established by the company and it was compulsory that all employees participate at a monthly charge of one dollar. At this nominal fee, each family was eligible for any medical service without additional cost. This welfare plan became known as the Miner's Fund and the following information was posted for the observance of all members:

RULES OF THE MINERS' FUND

This fund, instituted for the benefit of the residents of New Almaden, is established upon the following basis:

I

"Employees of the Quicksilver Mining Company, heads of families, and all other adults residing at New Almaden, each pay, monthly, into said Fund, the sum of One Dollar. The money so contributed is held by J. B. Randol, Trustee, to be paid out for the following purposes:

1. The salaries of a resident Physician, and of a Druggist, and for the purchase of medical supplies.
2. For relief of contributors, whom circumstances may entitle to the same, and for other contingent expenses.

II

Contributors are entitled, without further payment, to the attendance of the resident Physician for themselves and their immediate families (except that cases of confinement will be charged five dollars), and will be furnished with medicines prescribed by him, on payment of cost.

III

When the Fund is subject to any expense for relief of persons indigent, or otherwise — say, for medicines, nurses or supplies — it will be regarded in the nature of a gift, or as an advance, to be repaid, as the Trustee may decide to be just, considering the circumstances of each case.

IV

It is expressly agreed that when the resident Physician is called to attend any person not contributor to the Fund, that there shall be a charge of not less than Five Dollars for each visit to be paid into the Fund, and to be charged against and collected from the head of the house where such non-contributor may be living.

V

The Trustee serves without pay and in consideration thereof, it is understood that the foregoing rules and regulations will be observed by all persons interested therein; and it is expressly agreed that all sums due, or to become due, to the Fund by the contributors, or any of them, shall be a lien upon the property of the contributors at New Almaden, and

upon any money due, or to become due them, for wages from the Quicksilver Mining Company, which said Company is authorized to pay over to said Fund, without further notice.

J. B. Randol, Trustee
New Almaden, February, 1883

During the years of the flourishing mining camps throughout the West, many young doctors started their medical careers in an environment made up of the most challenging and diversified experiences They were general practitioners and had to serve every conceivable situation that had any bearing upon the health and welfare of the community.

In the settlements of Hacienda, Spanishtown and Englishtown, the inhabitants became the patients of the company doctor in residence, whose obligations and duties became a daily routine of dedication and service. This total dependency of the population, made the doctor's life an arduous schedule of appointments and calls at any time of the day and night. It was not an uncommon experience for the doctor to travel the winding paths on horseback in the darkness of night and inclement weather, to deliver a child or into the mine shaft for emergency treatment of an accident. It was also not uncommon to call the doctor for an ailing animal or as a dentist in extracting a tooth or lancing a gum in ulceration.

During the mining years of New Almaden, the inhabitants raised large families. The average was about four and went as high as eleven. In practically all cases of childbirth, the event took place in the home without benefit of hospital facilities. If they survived the first seven years, having avoided or recovered from the common childhood diseases, the outlook for survival was quite promising. William Bunney, a veteran blacksmith of the quicksilver days, once remarked:

"They raised a lot of kids on the Hill and they were all born in the home and as far as I know and I remember a lot of them, all the births were successful."

One of the common problems that faced the doctor, was the delayed time that people took to call for service. Many of the inhabitants were unaccustomed in seeking medical help for what they diagnosed as common ailments and they relied chiefly upon home-made remedies. In most of the homes, there was always the conspicuous health almanac and occasionally a book of remedies for self-doctoring. Old family prescriptions had been passed on for generations and there was always someone available to offer advice for the remedial treatment of aches, pains and illness. The survival

DR. S. E. WINN, a Doctor on Horseback

Dr. Winn was engaged by the mining company in 1879 to serve the people of Hacienda, Spanishtown and Englishtown. In Englishtown, the company provided an office, dispensary and house free of expense. During his ten years of service, Dr. Winn, traveled the rough terrain night or day with horse and buggy or on horseback, averaging about 5,000 calls a year. His interest in photography resulted in many fine pictures depicting the scenes of New Almaden.

of most patients seemed to substantiate the validity of the treatment. Many were guided by superstitions such as the dangers of night air and its cause of many afflictions. At certain times of the year, sulphur and molasses was a common concoction administered to all members of the household as a blood tonic and purifier. Young children were supposedly protected from certain disorders by a bag of asofetida suspended around the neck. This bag of ingredients consisted of a gum resin of several oriental plants of the carrot family, which in spite of a very offensive odor, was adopted as a common practice for the younger generation.

The greatest practice of home-made remedies took place during the winter season when colds laid siege to a large proportion. Every household was experienced in a variety of treatments that for the majority of the population were accepted practices. One of the common means of treating the patient was the application of the mustard plaster.

A portion of powdered mustard mixed with flour was made into a paste and placed between several layers of cloth. This was applied to the chest upon retiring to bed and kept on as long as possible. This mixture was quite strong and within a reasonable time, created an intense red coloring and burning of the skin. The purpose of the application was to stimulate circulation and give relief to the congested area. Also, in severe cases, the patient would be treated with a mustard foot bath. A portion of mustard was dissolved in a container of very warm water into which the patient placed his feet. The treatment lasted about twenty minutes during which time warmer water was added. The conclusion derived from this practice was that the blood circulation was stimulated for withdrawal from the congested area. For the treatment of coughs there was mixtures of varied ingredients which was a familiar product in every home. One common formula was a mixture of garlic, brown sugar, vinegar and honey added to a cup of warm water. In cases of sore throat, a small tube was made from paper and a portion of sulphur was placed in one end. Placing the tube in the mouth, the person administering the treatment would blow through the tube, spraying sulphur into the open throat. For the patient with fever, quinine was the most common and available drug. In cases of young children, paregoric, a camphorated tincture of opium was used in many situations of distress and a few drops added to a portion of water, was an accepted practice for the soothing of pain. The amount of dosage was determined chiefly by discretion.

In 1876, Dr. A. R. Randol, a brother of the manager, assumed the duties as the first resident doctor but after a few months of initiating the new plan, resigned to be succeeded by Dr. F. V. Hopkins, who carried on the work until 1879. During these few years, the people had found favor with the available health service and the doctor was accepted as an important member of the mining settlements. The health conditions showed improvement and the company benefited considerably by the reduction of worker absenteeism due to sickness.

Dr. S. W. Winn, a medic of some experience, arrived in September, 1879, and established residence on the Hill where he would spend ten years of dedicated service. The position was made more appealing by the company, in salary and facilities. Dr. Winn as a full time practitioner received $400 a month with house, horse and buggy. A dispensary with an office was constructed in Englishtown and, also, a small one in the Hacienda. The salary and expenses of the doctor were paid from the Miners' Fund.

With the population of some 1400, the demands upon the doctor continued to increase and the fulfillment of the many obligations were a constant challenge of stamina. For a great percentage of the population, the doctor had a very close, personal relationship and in many situations was the main source of information regarding special cases in need. He often made requisitions for certain supplies or provisions which were financed by the Miners' Fund. Such cases were common resulting from extended illnesses or accidents which deprived the worker of any income.

Much of the doctor's time was spent in attending a patient in the home which involved considerable travel to cover the three settlements with horse and buggy. For the minor or casual consultation, he was available at his office from 11 to 12 each morning excepting Sunday. At the Hacienda he, also, gave an office hour from 12:30 to 1:30 p.m., six days a week. In conjunction with the daily routine, there were always demands that could be anticipated at any time, for administering first aid to the victim of an accident in the deep working of the mine. The general routine of the doctor's life in the Cinnabar Hills left little time that was entirely free of demands.

A conservative accounting of his laborious practice estimated an average of 5,000 professional calls a year. Between the years 1874 and 1883, 84,000 cases were personally treated by the doctor in the home or at the dispensary. The total cost for service was $79,000 which was ably handled by the Miners' Fund. Along with the common ailments, of aches and pains, were afflictions of a more serious nature. The range of diseases which were recorded as prevalent in varying degrees consisted of Tuberculosis, Cholera Infantum, Meningitis, Capillary Bronchitis, Peritonitis, Erysipelas, Tetanus, Nephritis, Diptheria, Grippe, Cancer and Measles.

In 1889, Dr. Winn terminated his services and left the red dust paths that he had traveled for many a mile, for the Montana country where he would confine his abilities to private practice. His work in New Almaden had been a monumental contribution of service and the general welfare of his many patients had been greatly rewarded by his diligent and conscientious devotion. He had been a true practitioner of help and the day of departure was a sad parting of the ways as he left the many personal associations and the offspring that he had brought into the world.

Dr. J. Underwood Hall Jr., a young, unmarried, medic who was at the beginning of his career, arrived in New Almaden as the new replacement. He had been personally selected by James Randol

as the best qualified applicant available to assume the duties and carry on the excellent program to which the people had become accustomed. His first months of indoctrination to a remote mining camp and the development of a favorable rapport with a variety of clientele who still retained a certain partial attitude for their former practitioner, was a challenging situation for the young Dr. Hall. A letter from secretary Bulmore to J. B. Randol gives some indication of the health situation at this time.

New Almaden, Cal.
31 December, 1889

J. B. Randol, Esp.
Trustee, Miner's Fund
Dear Sir:

I have the honor to submit the Financial statement of the Miners' Fund for the year 1889. The net balance to the credit of the Fund is $7,185.18, whereas, at the same time last year, the amount was $7,252.54, showing a decrease of $67.35. Compared with the year 1888, our receipts at the mine have fallen off $239.35 (cause, decrease in population) and our expenses decreased $880.14. For this year the sales of drugs has exceeded the cash by $149.62 and in this cash is included $212.34 for vials, corks, boxes, utensils, surgical dressings, disinfectants, chlorate of potash and myrrh (for salivation cases), issued gratis.

Dr. S. E. Winn, our resident physician, retired during the year, after ten years of service. His successor, Dr. J. U. Hall Jr. who was appointed by you in September last, appears now to give satisfaction, and I feel assured has the confidence of our people. I say now, because when first appointed, there was some slight dissatisfaction, the appointment oweing to Dr. Hall's youth and single state.

The physician's report shows that they made 3499 professional visits, had 2040 office consultations, 25 obstetrics cases. These figures show an increase of work performed, when compared with previous year. The almost incessant rain for past two months, has caused considerable sickness, but previous to this, the health of the people has been fairly good and the schools have not been closed a day by reason of any epidemic amongst the children.

The Helping Hand Clubs at Hill and Hacienda, are in as much favor as ever and I am pleased to say the expenses to run them has decreased. 1888 — $466.12, 1889 — $278.05. The fame of these clubs has reached the East and we have had communications as to our workings, rules and fees.

Trusting that this report will meet with your approval,

I am, your obedient servant
Robert R. Bulmore, Secretary

After a certain period of orientation and acceptance by the settlements, Dr. Hall proved his qualifications and continued the fine program in which he contributed a quality standard of service. After five years, Dr. Hall resigned to engage in private practice in Santa Clara County and during his many years of service was highly regarded as a physician and surgeon. The following letter indicates the qualifications desired in a successor:

New Almaden
February 7, 1894

J. B. Randol, Esq.
Trustee, Miner's Fund

Dear Sir:

Enclosed, please find resignation of Dr. Hall, which I very much regret to receive and we must accept.

Will you kindly help us to obtain a gentleman to take his place. This is a difficult matter to do, as we require so much in one man — physician, surgeon and druggist and to properly fill the place, a man who has knowledge of Spanish, this latter not being important however.

Yours very respectfully
R. R. Bulmore, Secretary

Following the departure of Dr. Hall in 1894, the health program continued with doctors whose services were of short duration. Dr. Frank Lowell, Dr. A. M. Smith, Dr. W. T. Jamison and Dr. J. J. Kocher filled out the remaining years in serving the needs of the Almadeners.

With the health of the people receiving proper attention from professional treatment, James Randol expressed a strong interest in the social and recreational life of the settlements. His strong desire to build an ideal community, that in most cases was incongruous with most mining camps, was conscientiously expressed with personal consideration. He recognized the need for experiences and associations by which the people would mutually participate in activities of an aesthetic and recreational nature. In 1886, he promoted the construction of two buildings, one in Englishtown and the other in the Hacienda. The company assumed the financial cost of the project and also, furnished the interiors with adequate equipment and facilities. These buildings became identified as the Helping Hand Clubs and were available at any time by qualified residents.

During the year 1890, the settlements of Hacienda, Spanishtown and Englishtown were carefully screened and tabulated to constitute a census report with the following statitsics:

CENSUS REPORT AT NEW ALMADEN IN 1890

SPANISHTOWN

	Over 5 Years		5 Years and under		Total
	Male 282	Female 190	Male 47	Female 42	571
Number of deaths	4	6	7	2	19
Number of births					28

ENGLISHTOWN

	Male 283	Female 214	Male 39	Female 35	571
Number of deaths	1	2	2	1	6
Number of births					20

HACIENDA Spanish-American

	Male 24	Female 22	Male 4	Female 6	56
Anglo-American and others					
	Male 91	Female 57	Male 11	Female 8	167

DEATH RATE AT NEW ALMADEN IN 1890

	Per 1,000
Spanish-American	30.8
Anglo-American and other nationalities	8.1
Average death rate	18.5

BIRTH RATE AT NEW ALMADEN IN 1890

	Per 1,000
Spanish-American	45.4
Anglo-American and other nationalities	27.1
Average birth rate	35.4
Total existing population in 1890	1,413

By 1880, the mining activity had gained considerable momentum in further prospecting, development work and the construction of mine shafts. Three main shafts were in operation, namely, the Randol, Isabel and the Cora Blanca. The Randol, which was named for the manager, was the most prominent and through the years, became the greatest producer.

James Randol, who by this time was referred to as J. B., had reshaped not only the community life but had established the mining operation on the strictest routine of economical production. He carefully screened every detail for improving efficiency and the working conditions of the miners. Recognizing the importance of capable and dedicated employees, he placed special emphasis on obtaining the most efficient workers for specific types of work. Having observed the methods and efficiency of the Cornish miners, he encouraged their coming from Cornwall and western mining camps. There was also, a small minority of Irish, Swedes, Germans, Spaniards, Chilenos and native born Americans.

Following the Gold Rush, many Chinese arrived to accept any menial labor available. They followed the footsteps of the adventurous horde who worked the streams and gullies and were content to pick

This underground scene was taken at the 1500 foot level of the Randol Shaft. A load of rich cinnabar is on the way to the hoist. This phase of mining is called "tramming." Men engaged in this work were called "trammers" and in many cases were paid on the tonnage basis of ore loaded and delivered.

Mineros burrowing into the hard rock walls to bring out mercurial ore.

Great quantities of redwood timber was transported from the Santa Cruz mountains for the construction of tunnels into Mine Hill.

up what was left. A small group found their way to the Cinnabar Hills where they were employed on the dumps and about the furnaces. They were hardy, industrious and able to extend themselves in back breaking labor. Their pay was small but they were able to subsist on meager rations. They were tolerated rather than accepted and were often subjected to various pranks and unpleasant treatment. After 1880, the small Chinese camp was disbanded and they were released by the company.

With greatly improved conditions for living, working and the continued extension of development work, Randol was able to get the most experienced and competent workers in their respective areas. Good miners were a dedicated group and were attracted to the mines that offered the best wages and steady employemnt.

The New Almaden mines operated with two work shifts. The miners were underground from 7 a.m. to 5 p.m. with one hour for lunch. They were allowed one half hour for going to and from their place of work. The same hours were followed on the night shift. On Sunday and national holidays, the work was stopped in the mines and on the surface, except at the furnaces. Some workers were paid by the month, others by the day and for those underground the contract system was in practice. From the standpoint of economy and production, the contract system was the most advantageous for both the employee and the company.

It is a general assumption that the contract system had its origin in the mines of Cornwall and is thought to have had some bearing in the

Wal Malone, stage driver and teamster, transporting ore to the reduction works in 1892.

development of their skill and understanding to which they were held in high regard. The contract system motivated an incentive for the individual in a personal enterprise in which their abilities were rewarded according to their production. In general mine operation, the contract system dealt chiefly with the extraction of ore and the pay was based on tonnage.

Contracts were offered for the miner's bid and a notice was posted at the mine office several days in advance of acceptance. The information indicated the location, size of drift and number of men required. The men worked as a team and were well paired for certain methods and techniques. In a majority of cases the contract was awarded to the lowest qualified bidder. Under this system of operation, there was a natural selection and elimination of workers. The less competent were not able to compete with the more skilled and seasoned operators.

The systematic operation of Randol placed the general procedure into specific categories for the extraction of ore. Men were engaged in the following classifications on the Hill:

Yardage — This work required experienced and skillful miners. They were on a contract basis and were paid on the actual number of lineal yards extracted in a tunnel, drift or shaft.

Drillers — These workers were paid by the foot as measured by the length of the drill hole driven into an ore chamber. They worked under a shift boss who measured and reported their work.

The Tribute System — The contractors are paid for the amount of ore after it is cleaned at the planilla. He blasts and does the whole excavation of the ore.

Trammer or Tramming — These workers transported the ore in cars through the tunnels to the shaft where it was hoisted to the surface. They were paid according to the amount of tonnage delivered.

Skip-filling — The ore was transferred from the cars to the hoist. The workers were paid on the tonnage hoisted.

After the ore was carted to the planilla and sorted, it was then transported by wagon teams to the tramway where the cars were loaded for the descent to the furnace yard. Frank Bohlman, who operated a large livery stable in Hacienda, was the chief contractor of transportation for the company. With some 130 mules and horses in service, it was a common average to keep twenty teams actively engaged in hauling ore, lumber from the hills and miscellaneous freight to and from the mines. All of the industrial transportation was conducted by the contract method. Some thirty-five teamsters were employed in this work at a monthly salary of $35 with board.

The policies, established by the Quicksilver Mining Company concerning employment and the employees obligations, were specifically expressed by Randol in his rules and regulations. Selected excerpts from the Civil Code of California served as a basis for the items which would govern the new and old employees in the execution of their duties. The imposed rules and regulations made clear the relations and obligations of the employee and created a company autonomy for dealing with disciplinary measures and practices which violated current policies. It, also, served as a safe-guard for lawsuits or expenditures for which the company could avoid any liability.

33

In 1883, due to various unfavorable experiences, James Randol issued to all employees, the rules and regulations by which all working conditions would be governed. Also, in conjunction with the edict, the employee was enlightened with special excerpts from the Civil Code of California.

RULES AND REGULATIONS
of
THE QUICKSILVER MINING COMPANY
New Almaden, California

1st — The Company, its agents and employees, will not be bound to indemnify any employee for losses suffered by the latter in consequence of the ordinary risks of the business in which he is employed, nor in consequence of the negligence of another of its employees, unless the Company be proved to have neglected ordinary care in the selection of the culpable employee.

2nd — Each employee is required to perform his service in conformity with the usages of the mine and the works, unless otherwise directed by his superior officer.

3rd — Every employee is desired to use such skill as he possesses, so far as the same is required for the service in which he is engaged, and also use material and time in the most economical manner.

4th — No material nor tools will be allowed to any employee unless by permission from the office, and employees leaving must bring a receipt from their foreman in full for all tools used, or pay for them.

5th — Every employee will be held liable for all damage caused to the Company's property by his blunders, mistakes or carelessness, and will be paid for the value of such services only as properly rendered.

6th — The Company reserves the right to discharge an employee at any time in case of willful breach of duty, or in case of his habitual neglect of the rules, regulations and usages established for the welfare of all employees.

7th — Employees by the month must perform all the requirements of their respective service without charge for extra time, and such employees, in case of absence, must first obtain leave from their superior officer if they wish to retain their positions. If they desire to quit the Company's service, they must give one month's notice, or forfeit one month's pay.

8th — Regular pay-day will be on the last day, or the last Saturday of each month, as may be most convenient, when payment will be made for the preceding month. Payments will not be made at any other time, except it shall suit the Company to do so. All payments are due at the Company's office in New Almaden, and at no other place.

9th — It is required of every foreman to call attention of each and every employee to the foregoing rules before he begins work for the Company, and it is understood that in consideration of their employment each and every employee agrees to recognize the foregoing rules as a specific contract between employer and employee, and to faithfully abide thereby.

THE QUICKSILVER MINING COMPANY
J. B. Randol, Manager
February, 1883

EXTRACTS FROM
CIVIL CODE OF CALIFORNIA
Division III, Part IV, Title VI, Chapter I

ARTICLE II

Sec. 1970 — An Employer is not bound to indemnify his employee for losses suffered by the latter in consequence of the ordinary risks of the business in which he is employed, nor in consequence of the negligence of another person employed by the same employer in the same general business, unless he has neglected to use ordinary care in the selection of the culpable employee.

ARTICLE III

Sec. 1981 — An employee must substantially comply with all the directions of his employer concerning the service on which he is engaged, except where such obedience is impossible or unlawful, or would impose new or unreasonable burdens upon the employee.

Sec. 1982 — An employee must perform his services in conformity to the usage of the place of performance, unless otherwise directed by his employer, or unless it is impracticable, or manifestly injurious to his employer to do so.

Sec. 1984 — An employee is always bound to use such skill as he possesses, so far as the same is required for the service specified.

Sec. 1988 — An employee who has any business to transact on his own account similar to that instructed to him by his employer must always give the latter the preference.

Sec. 1990 — An employee who is guilty of a culpable degree of negligence is liable to his employer for the damage thereby caused to the latter; and the employer is liable to him, if the service is not gratuitous, for the value of such services only as are properly rendered.

The Buena Vista pump engine was the largest in the mines. This power plant with the giant flywheel measuring 24 feet in diameter and weighing 25 tons operated the Cornish pump. The excellent functioning of this equipment served well its purpose of extracting water from the lower depths. The pump made 8 double strokes a minute and brought to the surface 23,400 gallons of water an hour.

The sturdy blacksmith crew who were engaged at the Randol shaft.

THE CAPTAIN JAMES HARRY SHAFT

This exploratory project was started in 1893 and was the last major shaft of the Quicksilver Mining Company. It was named for the man who had spent many active years as a mining captain in the general underground workings of the New Almaden mines. The shaft terminated at a depth of 800 feet.

ARTICLE IV

Sec. 2000 — An employment, even for a specified term, may be terminated at any time by his employer, in case of any willful breach of duty by the employee in the course of his employment, or in case of his habitual neglect of his duty or continued incapacity to perform it.

Sec. 2002 — An employee dismissed by his employer for good cause, is not entitled to any compensation for services rendered since the last upon which a payment became due to him under the contract.

The inexperience of James Randol, in the technicalities of mining was not obvious because of the efficient and experienced personnel that he was most fortunate to engage during his years of management. During his first twelve years, he relied entirely upon the mining captains who were in complete charge of the underground operations. A mining captain was first, a miner who had experienced many situations of mine operation and

possessed the intelligence and ability to direct and handle working crews in any given situation. Randol's approach, to the daily schedule was almost academic, but his ability and keen mind made it possible to visualize and understand the existing problems from charts and maps described by his aids. According to rumor, Randol was never seen entering or leaving a shaft or tunnel during his years as general manager.

As the New Almaden mines maintained production of quicksilver, the market demands were ever increasing. Competition from within and abroad stimulated greater production. Certain circumstances indicated to Randol that there had been quicksilver sales under the disguise as New Almaden production. Realizing the challenge of the competitors, Randol took special precautions to safeguard against the unscrupulous producer. He prepared a statement which was posted and published as a protection against the unethical infringement and also to identify New Almaden quicksilver by the legal procedure of a patented

Trade Mark. The following statement was submitted to the United States Patent Office on September 29, 1873 and the Trade Mark was registered on October 14, 1873:

"Specification describing Trade Mark, used by the Quicksilver Mining Company, a Company charted by the State of New York and working quicksilver mines in New Almaden, Santa Clara County, State of California, for quicksilver flasks."

"Our Trade Mark consists of a letter (A) straddling a circle. This letter has generally been painted upon the upper end of each flask, bottle or jar in which quicksilver is contained, its usual position being such that the opening or mouth, in the upper end of the flask, bottle or jar, will be between the spreading angular sides of the letter, the cross-mark of the letter passing along close to the edge of the hole.

This Trade Mark we have used in our business for ten years last past. The particular goods upon which we have used it are quicksilver flasks, and it has always been applied as above described."

The Quicksilver Mining Company
J. B. Randol, Manager

In conjunction with the established trade mark and its significance publicized, Randol followed further with additional information to alert purchasers of quicksilver regarding the New Almaden product.

NEW ALMADEN QUICKSILVER TRADE (A) MARK

"The well-known full weight and superior quality of the quicksilver produced at New Almaden Mines, having induced certain unscrupulous persons to offer their inferior productions in flasks having our Trade Mark A, notice is given to consumers and shippers that quicksilver (A) brand, is guaranteed weight, can be purchased only from Thomas Bell, San Francisco, or his duly appointed sub-agent."

J. B. Randol, Manager
New Almaden
April 15, 1875

In San Francisco, the sales of New Almaden quicksilver was handled by Thomas Bell. In the early days, when the New Almaden mines were initiated into productivity by the Barron, Forbes Company, Thomas Bell had been associated personally and professionally with the firm. This adventuresome individual arrived following the Gold Rush and through his talents and initiative, established his status within the circles of high finance, mining and railroads. Also, in his career were experiences in treasure hunting, land manipulating, and a fair degree of scandal, notoriety and lawsuits. The colorful career of Thomas Bell has been interestingly portrayed by Helen Holeridge in her book, "Mammy Pleasant's Partner."

As a representative for the Quicksilver Mining Company, Thomas Bell advertised regularly in the San Francisco papers which was cautioned with the adopted Trade Mark (A). The following statement appeared at regular intervals:

"This Trade Mark is also registered in the office of the Secretary of State, Sacramento, California, and all producers and dealers in quicksilver are cautioned not to use the said Trade Mark."

For New Almaden quicksilver apply to

Thomas Bell, Sole Agent
Over Bank of California
San Francisco

One of the most challenging and unpleasant experiences of Randol during his management, was the accusation brought about by Frank J. Sullivan, a defeated candidate for Congress. It was Sullivan's contention that the voters of New Almaden were strongly influenced and co-erced to vote according to the dictates of the manager, James B. Randol which resulted in losing the miner's vote and election to office.

Following the general election of November 2, 1886, in which Charles N. Felton defeated Frank J. Sullivan for Congress, in the Fifth Congressional District, the inhabitants of the Cinnabar Hills became subjected to much publicity by the losing candidate, as he contested the results of the election. The issue was taken to court at New Almaden on April 19, 1887. The following article, which appeared in the San Jose Daily Mercury at the close of the proceedings, gives an interesting summary of the affair:

CONTESTED ELECTION

"At the general election held in California, November 2, 1886, Hon. Charles N. Felton and Frank J. Sullivan were rival candidates for election as members of Congress in the Fifth Congressional District. The official returns gave to Hon. Charles N. Felton a plurality of 188 votes, and the certificate of election duly issued to him.

THE SANTA ISABEL SHAFT

This shaft was started in 1877 and reached a depth of 2300 feet. It was the deepest shaft extending from ground surface and during its years of activity, was an excellent producer. An added feature in this shaft was the discovery in 1894, of the first Carbon Dioxide gas in California. The discovery area was sealed and the gas taken out for market use. In the deep laterals, mules were used for tramming and were stabled underground.

THE VICTORIA SHAFT

One of the later mines located near Englishtown which failed to develop with any great promise. It reached a depth of 1000 feet.

Sullivan afterwards inaugurated a contest for the seat. In this contest, the vote of the New Almaden precinct became a very conspicuous part. The majority against Sullivan in that precinct was 164, so that if Sullivan could have that precinct thrown out, or the votes for Mr. Felton counted for Mr. Sullivan, he might reasonably expect to unseat Mr. Felton and fill the place himself.

The New Almaden precinct is the seat of the great New Almaden Quicksilver Mine, which for many years has been the principal quicksilver producer of the American Continent, and employs about 500 men in all capacities, of whom 156 were voters, and voted at the election in 1886. The total vote of the precinct was 256.

The grounds of the attack upon the vote of New Almaden precinct by Sullivan were, that the voters of the precinct were co-erced, intimidated, and held in a state of peonage or slavery and compelled to vote for Mr. Felton against their will.

In support of this he attacked the method of conducting the mine, the system of trade at the mine, and the general condition of the miners; claiming that those who did not vote as pleased the management of the mine would be discharged.

On behalf of Sullivan the testimony relating to Almaden was almost entirely hearsay so far as it could be in any sense pertinent to the issue; being generally a statement by witnesses that they "had heard," or "it was generally understood" that so and so had happened at Almaden.

On behalf of Mr. Felton, a large amount of evidence was taken, very much of it at New Almaden. Of the 256 persons who had voted in that precinct in 1886, 243 were personally examined, the other 13 being dead or having removed to a distance, or for other reasons their attendance as witnesses being impracticable. Of the number examined there was not one but what testified in the clearest terms, that for more than fifteen years past the subject of voting, or candidates, or elections had never been mentioned to any employee of the Company by any one in any way connected with the management of the mine; that the Superintendent of the mine was the only person in the precinct who voted the American ticket in 1886.

That the voting employees of the mine consist very largely of natives of Mexico and Cornwall, England, who here, as elsewhere in California, vote the Republican ticket almost exclusively. It was, also, shown that there were in the employ of the Company and always had been, several Democrats who were always outspoken in favor of that party and its candidates and who had never had their relations with the Company affected by that fact.

That the vote for Sullivan had been about the average vote given to his party, but that some candidates of the Democratic party, on account of acquaintance or personal popularity had received a larger vote.

As to the system of trade at the mine, it was proven that at the mine were two extensive and well stocked general merchandise stores, the proprietors of which are in no way connected with the mine, or its management, or officers. That the employees of the Quicksilver Mining Co. have facilities for purchasing goods at these stores upon a system of credits or orders, known as boletos, issued by the stores, and based upon wages earned, but not yet due to the employees.

The social and financial condition of the employees of the company, was shown in considerable detail. It appearing that they are better fed, better paid, better clothed and better housed, than any other laboring class in the State. A large collection of photographic views was introduced in evidence for the purpose of showing the homes, surroundings and conveniences supplied for the laboring people. It was also shown that a majority of the miners had money on deposit in savings banks, and owned stocks in dividend paying corporations. That the management, and especially J. B. Randol, the manager of the properties of The Quicksilver Mining Company, had devoted much time and attention to the elevation of the people living at the mine. They were supplied with handsome churches, schools, clubs, lodges, and societies all well sustained and flourishing. That a very large number of daily and other periodicals representing all political views were taken regularly by the employees at the mine.

That nearly all the work at the mine is let and carried on by the contract system. The work being let in small monthly contracts (employing from two to twelve men each) to the lowest bidders, in a competition open to all, no favors being dependent upon politics or anything else.

39

Hoisting engines at the Buena Vista shaft.

Miners coming to the surface in the compact cage after a 10 hour shift deep in the drifts of Mine Hill.

That not a man in the employee of the company, who had voted previous to coming here, had changed his party fealty after coming to Almaden.

In fact the investigation showed that there was no possible ground for the assault upon this precinct, or this mine or its management, unless it be found in the undisputed statement of one of Sullivan's witnesses, who testified that about the time of the initiation of the contest Sullivan told him in his private room at the La Molle House in San Jose, that "he intended by this contest to cast a shadow upon the entire Congressional District," and then rely upon a Democratic majority in the House of Representatives, to give to him Mr. Felton's seat."

The conclusion of the contested election hearing brought little satisfaction to Frank Sullivan. Those voting from Spanishtown, Englishtown and the Hacienda showed a great percentage to be Republicans and had voted for their party candidate, Charles N. Felton. The general information made public from the testimony of 243 persons, clearly indicated that the relationship between management and labor was highly satisfactory. There was no evidence to show that the personal life of the employees was influenced, co-erced or controlled by the administration.

During the half century of quicksilver production on Mine Hill, the way of life and the economy of the mines and people carried on with little change. Wages and prices in general, followed closely the prevailing conditions throughout California. On wages ranging from $40 to $100 a month, the people

THE BUENA VISTA SHAFT

Work started on this shaft in 1882 and continued to a depth of 2300 feet. The massive granite block foundation was the most substantial and elaborate of all the shaft buildings. In this structure were housed the largest installation of pumps and hoisting equipment.

The St. George shaft was started in 1877 and developed to a depth of 1200 feet.

JAMES B. RANDOL SHAFT

This shaft was commenced in June, 1871 on a spur of Mine Hill, and continued as a great producer until 1896. The shaft and laterals uncovered ore bodies 100 to 300 feet wide and 12 to 20 feet in thickness. The Randol proved to be abundantly rich and produced over $10,000,000 in quicksilver during its years of operation. It terminated any further development at the 1800 foot level.

THE AMERICAN SHAFT

This mine, previous to 1864, called the "Bull Run," was opened in September, 1863. The workings were plagued with hard rock and excessive water. In 1885 a steam hoisting works and pump were installed which functioned until July, 1888, when the underground area was subjected to great flooding, submerging the equipment and causing an abandonment of the mine. The shaft had reached a depth of 700 feet.

in New Almaden were able to maintain a favorable standard of living. In 1889, the company report gave an itemized expenditure for labor showing the following classifications. Those workers engaged in contract work were not included.

THE QUICKSILVER MINING COMPANY
New Almaden, California

WAGE SCALE 1889

Mine Employees — Hill Mines

	Monthly Rate
MACHINIST	$ 100
MACHINIST HELPER	60-75
ENGINE DRIVERS	70-80
FIREMAN	40-60
BLACKSMITH	45-60
BLACKSMITH HELPER	30-35
PUMPMEN	70-90
SHAFTMEN	71
BLASTERS	71

	Daily Wage
BOILERMAKERS	$ 2.00-2.50
TIMBERMEN	2.00-2.50
CARPENTERS	3.00
SURFACE LABORERS	1.50-2.00
ORE CLEANERS	1.75-2.25
LABORERS IN LABORES	1.50-2.00
TRAMMERS	1.75-2.25
SKIP FILLERS	2.00
BOYS	1.00-1.50

THE HACIENDA — REDUCTION WORKS

	Monthly Rate
FURNACE FOREMAN	$ 100
WEIGHERS	70
MACHINIST	70-100
MASON	150

	Daily Wage
BLACKSMITH	$ 3.00
BLACKSMITH HELPER	1.25-2.00
CARPENTERS	3.00
ENGINE DRIVERS	2.00-2.50
LABORERS	2.00-2.25
FURNACEMEN	2.50
SOOTMEN	2.10
TEAMSTER (2 horses)	4.00
TEAMSTER (4 horses)	6.00

James B. Randol closed his career at the New Almaden mines in 1892, at which time he had concluded twenty-two years as a superb manager in developing the Cinnabar Hills to its greatest prestige as one of the world's great quicksilver mines. His years of service were clearly expressed in his humanitarian motives for developing an ideal community for the employees that they might enjoy the amenities generally uncommon in most mining camps. His high standards of operation were followed by his subordinates in their execution of company policies. From the time of arrival in 1870, when production had declined to less than 15,000 flasks a year and the ore bodies were nearing exhaustion, he fulfilled all of his expectations in building the industry and its community. The mines attained the peak of their development under his management and the exploitation of their resources was consistently maintained through what proved to be the bonanza years of the New Almaden mines. By 1892, the available ore bodies had yielded of their treasure and deeper exploration was beyond the facilities and capital to continue. The situation now, was quite similar to that when Randol arrived in 1870. Non-productive labor, shafts that ended in caverns of exhausted ore and quicksliver production being down to 13,000 flasks a year had considerable bearing on bringing about the end of the Randol regime. While the company had not paid dividends for many years, the capital resources were insufficient to continue further work on a speculative basis. It was the consensus of opinion that Mine Hill had run its course. The situation was apparent to James Randol that the great days of quicksilver mining in New Almaden were over and continued operation was possible only by drastic reduction of cost. Work crews were reduced and many employees of long standing were dismissed from the payroll.

In 1877, Hennen Jennings, a Harvard graduate, arrived at New Almaden where he was employed as an assistant surveyor. He had spent several years in the Mother Lode mines getting his first experience in mining operations. He had been recommended by J. Ross Browne who had formerly done surveying work for the Quicksilver Mining Company. After one year as an assistant, Jennings was appointed chief surveyor and continued in this capacity until 1883 when James Randol advanced him to superintendent. It was of vital importance under the prevailing conditions that exploratory work be extended deeper into Mine Hill. Such a project required a person of specific qualifications and experience. The operation in promoting the first deep exploration and development of new ore bodies became the primary objective of the new superintendent. The successful results of this venture clearly indicated that the appointment of Jennings for such a task came at the most opportune time. His well-conceived plans and strategy of direction prolonged the life of mining activity in New Almaden and a critical appraisal of his accomplishments established his prestige as one of the outstanding superintendents.

ROBERT BULMORE

Mr. Bulmore was a Londoner whose early years were spent as a bank accountant. He became an employee for the Commercial Bank of India in Hong Kong following several years experience in India. He later sailed for California and New Almaden, where he was engaged by J. B. Randol as cashier and foreman at the Hacienda office in September, 1878. At the retirement of Randol, Bulmore was appointed General Agent and Pacific Coast representative for the company. This position he held until December, 1899, when the same was abolished due to the decline in quicksilver production. The Bulmore family were the last official occupants of the Casa Grande.

In 1887, Hennen Jennings had reached the termination of any further exploratory work and the future outlook for maintaining sufficient ore production became more speculative. Having been alert to new opportunities, Hennen Jennings resigned and left New Almaden to become superintendent of large scale mining in Venezuela. His position was filled by Col. F. Von Leicht.

During the last two years of Randol's administration, he resigned from active duty and served as General Agent. He relinquished the directing of mining operations to his superintendent, Col. F. Von Leicht, who later resigned and was replaced by Captain James Harry. When Randol officially left the company in March, 1892, he recommended for his replacement, Robert Bulmore, the company cashier since 1878, to take charge of the general business on the Pacific Coast. Captain Harry directed the mining activity until his death in 1895. His position was filled by Charles Derby, who had been the company surveyor.

James B. Randol left New Almaden to promote a newly discovered quicksilver mine in Northern California where his stay was of short duration. He returned to his native state, New York, where he died in December, 1903, at the age of 67. During his years in Santa Clara County, he had acquired some 600 acres of land, invested wisely in business transactions and was a leading stockholder of the prominent Vendome Hotel in San Jose. His personality, abilities and conservative manner left a lasting impression with the "Almadeners." They remembered his charitable attitude and the enthusiasm he expressed for improving the environment of New Almaden. In the true spirit of his time, he represented in every respect, the leading character, "Mr. Almaden."

CHAPTER 5

SPANISHTOWN

A Colorful Setting on the Slopes of Deep Gulch

Looking upward from the base of an open ravine, where steep irregular slopes were once the scene of a picturesque settlement of scattered homes, nature in its rampant pattern of foliage, has removed all traces of identification. Secluded in the mass of chaparral and wild broom, may still be seen some of the excavated plots where once stood the home of a Mexican miner. The winding footpaths and roads through the area, worn deep in the rocky, red soil with fifty years of habitation, are faintly visible to the observing eye. On the open ridge at the top, which was once the plaza with its Catholic church, cantina and schoolhouse, is now a scene of grazing cattle wandering leisurely over the terrain of sparse, stubby grass. Above this pastoral scene on a high knoll adjoining the ridge is the site of the old cemetery, completely covered with heavy growth. The picket fence and crude crosses have long been missing that would identify the people who have been resting in this tranquil solitude of desolation. Long forgotten and unattended, the little cemetery has become obliterated by the care-free pattern of nature's designing ways. And, in the quietude of this setting overlooking the wide ravine that was called Deep Gulch, there is little to disturb the stillness except the rustle of leaves, the song of a bird, or the barking of the ground squirrel annoyed by the footsteps of a stranger.

Deep Gulch, situated at the base of Mine Hill, had its origin about 1850. The early inhabitants were chiefly the workers who were transported from Mexico by the Barron, Forbes Company. Spanish speaking natives of the area gradually increased the population to the largest of the three settlements. As the entire village was closely associated with the Spanish language and only a small minority being able to converse in English, the settlement in Deep Gulch soon became known as Spanishtown.

Like the facilities in Englishtown, everything was company owned and operated with the same authority. House rent and other utilities were offered at the same rate. Many of the cottages that were situated on favorable ground were em-bellished with a white fence, vegetable and flower gardens. Wherever space was adequate, the Mexican yard would be a place for raising chickens, keeping a cow, mule or burro. Dogs were, also, a representative group in the population. Many of the homes were less fortunate in nature's facilities and clung to rocky slopes, accessible only by a narrow footpath.

Spanishtown, with its houses scattered at random, was in an open exposure to the elements of long, warm summer days and winter months of wind and rain. Heavy rain of any duration created eroding rivulets down the slopes and paths and on through the gullies to the Los Alamitos creek. This homogeneous community expressed an atmosphere of care-free living. Within a short space of time, this bizarre and desolate setting was transformed into a colorful and harmonious environment. The daily life of the people was typically expressed in the customary and traditional manner of their native land. Their unrestrained mode of living varied little except for occassions of celebration and festivity. While neighborly and congenial, they were not too receptive to other groups because of their meager understanding of English. Mary Halleck Foote expressed this observation after her visit in 1867:

'The Mexican camp has little of that bustling energy which belongs to its neighbor. It wakes up slowly in the morning, — especially if the morning be cold, — and lounges abroad on moonlight mights, when guitar tinklings sound from the shadowy vine-flecked porches. The barest little cabin has its porch, its climbing vines and shelf of carefully tended plants. Dark-eyed women sit on the doorsteps in the sun braiding a child's hair or chattering to a neighbor, who leans against the door-post with a baby half hidden in the folds of her shawl. They walk up and down the hilly street, letting their gowns trail in the dust, their heads enveloped in a shawl, one end of which is turned up over the shoulder; the smooth sliding step corresponds with the accent in speaking. In passing, they look at you with a slow, grave stare, like that of a child."

This photo shows the main area of SPANISHTOWN at the top of Deep Gulch. Painted cottages with fences and gardens created a feeling of permanency. On the crest of the hill stands the Catholic church etched against the background of the Santa Cruz mountains.

SPANISHTOWN IN 1890

When New Almaden mining began in 1847, most of the workers were transported from Mexico. On the steep slopes which converged into a deep ravine, the settlement that became Spanishtown, had its beginning. Winding foot paths were the only accessible means to many of the homes that clung to the slopes.

The upper section of Spanishtown, showing the Catholic church at the left and the schoolhouse near center.

To the casual observer in the early days of Spanishtown, the scene depicted the typical Mexican Aldea. The majority of the population was married and in their younger years. The adventurous spirit of leaving their native soil for a new environment was a rewarding experience, for life in Deep Gulch was an orientation to a higher standard of living. Up and down the narrow paths could be seen any day, women going to the church or to the store in Englishtown, draped in colorful shawls that covered the head and shoulders. Several days each week, the panadero, astride his mule with two large covered baskets strapped on the back, plodded the dusty course, delivering bread to his many customers. Over the ridge from near-by springs, young Mexican boys and their burros with two small barrels strapped over the back, delivered water to many of the homes. This means of water supply was a common practice for many years before the development of a piped water system by the company. During October and November, the Mexican and his mule were daily engaged in transporting stove wood which was used in every home. The sturdy mule served well the demands with many trips over the hills, from dawn to dusk. A conspicuous scene of activity was the large planilla, situated on the northern slope of Mine Hill. Day and night the ore cars emerged from the tunnel and unloaded their contents in this large shed where laborers were continuously engaged in cleaning and sorting the ore.

During the long summer season, Deep Gulch maintained a warm temperature. An occasional breeze in the late afternoon, from the bay area and the coastal mountains, alleviated the monotony of the summer heat. Dusk came early to Spanishtown as the sun set behind the high western hills, creating a feeling of coolness at the twilight hours. And into the late hours of dusk during the summer months, the people left their sun-baked houses to wander the paths, visit their neighbors or congregate for relaxing hours on the front porch. The children rode their burros up and down the dusty paths and the voices of the people, accompanied by the singing strings of the guitar, the laughter of the children and barking of dogs, created a spirit of compatible association.

Life in the compact settlement of Spanishtown, with each breaking of dawn was a casual expression of complacency day by day. "Quien canta, su mal espanta" (a cheerful spirit lessens many troubles), was the antidote to any situation of adversity. "Manana sera otra dia" (tomorrow is another day), which came as a matter of course and with little anticipation except for special occasions. Probably the most important event that created a change in the demeanor of the population was "Dia de Raya" (the day of plenty). This day came on the last of the month at which time, the mineros assembled at the mine office on the Hill to receive their pay envelope of silver dollars. Lacking in frugality and

MINE HILL

Rising above Spanishtown and Englishtown, was Mine Hill, 1755 feet above sea level. On the peak was a surveyor's monument from which all mine levels were calculated as to their depths. This photo shows the Main Tunnel and the big Planilla extending into Spanishtown. Authorities have stipulated that in this area, the first mining activity had its beginning in 1845.

(Watkins Photo)

THE BIG PLANILLA

The Planilla was the sorting shed where large crews of Mexican laborers segregated the ore according to quality and broke it down to the proper size for the furnaces. This scene of activity was in the upper area of Deep Gulch. The background shows a portion of Mine Hill and a small number of the miner's cottages. This is one of the earlier scenes taken in the 1870's.

DELIVERING STOVE WOOD

During the Fall season many of the young Mexican boys with their burros, delivered stove wood from the nearby hills to the cottages in the mining settlement. These sturdy animals carried two large bundles that averaged about 300 pounds for a distance of several miles. The crude cinching of the load with ropes kept the hides of these animals in continuous abrasion.

ERENCO GONZALES

One of the few pioneers still on the job in the 1890's that had come from Mexico with the first group brought to New Almaden by the Barron, Forbes Company in 1848. Gonzales was known as a Peon de Estribo or footman in the mines.

with little concern for the coming days, wages were spent with abandon. Senoritas arrived from San Jose to participate in the festivities of their friends who offered special treats with their traditional recipes. Throughout the village during the evening hours, Deep Gulch echoed with music and singing of Mexican songs. The company attempted to control these affairs, particularly the entry of unauthorized visitors, but were not always too successful. Occasionally, the celebration of Dia de Raya developed into a riotous affair but such an event was short-lived, for the silver dollars were soon spent as the local merchants absorbed most of the payroll. This was the cycle of living in Spanishtown, so spontaneously expressed by people who enjoyed their way of life. Their extemporaneous indulgence was rewarding for the laborious days in the depths of Mine Hill. The lean days that would follow were of little concern; Dia de Raya would come again.

Tradition, custom and religious functions were the most obvious factors that influenced the individual expression of the Spanishtown inhabitants. A

49

The rustic and partially whitewashed cabins of the Mexican miners located in the upper section of Spanishtown.
(Watkins Photo)

EDUCATION IN DEEP GULCH

This one-room schoolhouse was located near the plaza and served to orient the Mexican children to the 3 R's. The great percentage of the homes spoke only the Spanish language which made the teacher's chores much more difficult. After four grades the children went over the ridge to Englishtown to complete the Upper Grade work. The average enrollment was about 35 and the teacher received $70 a month. By the turn of the century the little schoolhouse was vacated for lack of students.

A view of Mine Hill and a part of Spanishtown. At the base is the San Francisco Tunnel. On the hill may be seen the extensive work of open-cut mining.

dedication to their racial inheritance was a motivating force in establishing a way of life in a new environment. Their complacent attitude was a consistent pattern of daily acceptance in the ups and downs of pleasure or adversity.

As days of special significance approached, the tempo for jubilation increased. Special events were anticipated with great interest and the energetic indulgence in any festivity or celebration was an individualistic expression in the most congenial and cooperative manner. One of the first festive occasions of the year was the Cascarone Ball, which took place on the Saturday evening before Ash Wednesday. This was an occasion of considerable concern, particularly for the women, whose initiative and ingenuity would be fully expressed in the details of preparation. This colorful and spirited event was a traditional custom of the Mexican people and preparations began far in advance. Each home became a busy workshop for the senoras and senoritas as they designed a new costume and created decorations to elaborate the fandango.

The most important activity, engaged in by the women, was the making of Cascarones. This Spanish term applied to the product created at the discretion of the individual in color and content. After selecting a required number of eggs, a small opening was made at the end and the contents emptied. The shells were set aside for drying and the task of cutting quantities of colored paper into confetti size began. By the following day, the shells were ready and each was solidly filled and the open end sealed. A touch of color or design to the outer surface, and a small basket of Cascarones awaited the hour of activity. While the women were chiefly the contributors to this detail of preparation, men in a prankster mood, quite often added their own inventions in which the Cascarone might be filled with water, perfume, flour or birdshot. It was generally anticipated by most of the participants that they would experience the unusual.

On the evening of this festive fandango, the large planilla that projected from Mine Hill into Spanishtown had been made ready for the colorful gathering. This open shed, encrusted with dust and piles

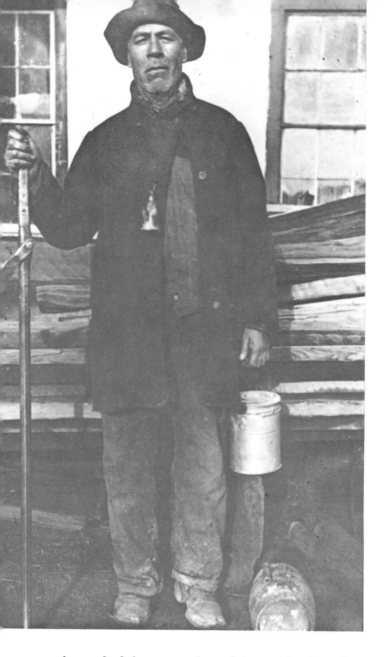

With the advent of the Barron, Forbes Company from Tepic, Mexico, came also the Mexican miner and laborer who worked in the mines in their native Mexico. California natives of Spanish descent were also at hand to promote the first development and mining operation. This picture exemplifies the typical miner from Spanishtown. While the Mexican miner proved capable in physical stamina for the many chores in mining activity, he seemed at his best in tracing and discovering ore bodies. A good Mexican miner was said to have a keen nose for pay-dirt.

---

that the dancing was done with corner partners instead of opposites. The music consisted mostly of waltz selections and Mexican folk tunes. Castanets and tambourines added that important touch to the rhythm of their native music. As the dancers swung into action, the people became more expressive in their feeling of gaiety. At opportune moments, the Cascarones began breaking on the heads of partners and others near at hand. As the music changed at regular intervals, the dancers circulated to select a new partner. Cascarones in great numbers were scattering their contents from the forceful contact with many heads. The humorous part of the affair now became evident for the individuals who were recipients of the unusual Cascarone. The handiwork of the prankster was ever conspicuous and was accepted as a customary surprise to be anticipated. The Bail de Carnaval ran its course to the depletion of the Cascarones and with a "buenos noches," "hasta la vista," the group gathered their lanterns and returned to their homes. While the Cascarone Ball was a festive expression for personal enjoyment, it was, also, of symbolic significance for the curtailment of certain indulgences befor the advent of Lent.

Another traditional event that attracted full community participation, with its routine of improvised humor, took place on Wednesday before Good Friday. This humorous dramatization performed in a ritualistic manner with a symbolic innovation was called the Colgante de Judas, or in the colloquial meaning, the Hanging of Judas. The affair was planned and conducted by a selected group who assumed the initiative of handling the details and preparation. Several would construct a gallows in the center of the plaza. A capable individual would create a life-sized effigy to represent the leading character, Judas. Some of the group were delegated to solicit the houses for nondescript articles to be used in the ceremony.

On the morning of the ceremony, the leaders of the affair assembled at the plaza for the start of the procession. The effigy, Judas, was carefully propped upright in a small cart attached to a burro. The procession started on a circuitous

of ore, had been transformed into a facility of appropriateness befitting this annual turnout of village celebrants. A variety of decorations and kerosene lanterns hanging on the supporting timbers of the shed created an atmosphere in harmony with the occasion. The preliminary hours, which would evolve into a climactic situation at a given time, were spent in dancing, singing, refreshments and socializing with one's neighbors. The music was furnished by local musicians in the community and quite often was augmented by outside talent.

At the arrival of the midnight hour, the gathering was informed to make ready for the culminating event which was called the Bail de Carnaval. The senoras and senoritas assembled with their little basket of Cascarones. The leader of the affair organized the participants into groups of four at designated places on the floor. The dance, as conducted, was quite similar to the Lancers except

This humorous dramatization of a symbolic and ritualistic nature was called by the Spanish speaking people, the Colgante de Judas, or the hanging of Judas. This rare photo taken in the 1880's, shows the setting for the culmination of the ceremony which was preceded by a colorful procession and the gathering of the inhabitants for participation in this annual affair.

journey through the village where waiting people along the route joined ranks. The journey was completed as the procession arrived at the Plaza and Judas was deposited under the gallows. The suspended rope with noose was attached around the neck and Judas was held in a standing position. A leader for the ceremony was selected for his ability to inject wit and humor into a improvised routine. A statement, in the form of a will, had been prepared which was to be read to the gathering. The large collection of articles at hand represented the earthly belongings of Judas and were to be disposed of to various individuals according to his wishes. All of the gathering were to be recipients of a gift which was pre-planned for its ridiculous and humorous aspects.

As the formalities came to a close, the hour of doom for Judas was near at hand. In the construction of the effigy, a double compartment was arranged in the torso. The upper part was filled with firecrackers and gunpowder while the lower part would, on the occasion, contain a cat. The leader indicated to the crowd that the hanging

was ready to take place. He advanced to the effigy and lit the fuse which projected from the torso. His assistant, at the side of the gallows with rope in hand, waited for the first signs of combustion. As the explosive contents became active, he pulled the rope, lifting Judas to a suspended position, several feet off the ground. The leader, who held a wire attached to the lower compartment, pulled off the opening which made an exit for the cat, as the explosive interior of Judas was in full force. The incarceration and release of the cat symbolized the departure of the evil spirit from the earthly remains of one who was being consumed in flames. As the effigy became a small deposit of ashes, the ceremony came to a close. The articles which were bequeathed to the crowd were returned to the donor, if requested.

The 3rd of May was the day designated to express a custom of a more serious nature. This day was referred to as the Dia de la Santa Cruz, which was the day of the holy cross. Each worker constructed for himself a small wooden cross. On the day of the ritual, the workers and their families

THE TOP OF THE RIDGE ABOVE DEEP GULCH

This scene near the plaza shows the Butcher Shop at the left. The loaded wagon is a dealer in second-hand materials.
Photo taken about 1885.

A HOME OF THE MEXICAN MINER 1885

The redwood dwellings that accommodated the miners on the Hill, were built, owned and operated by the mining company. These buildings are typical examples and offered modest accommodations for a monthly rental of about $5. On the knoll in the background, is the Mexican cemetery. Today this site is covered with growth and little remains to identify the resting place of many who came to stay.

THE WASHINGTON SHAFT

This shaft was started in 1881 and originally was called the Garfield. Following the assassination of President Garfield, the name was changed to Washington. The shaft was sunk to the 1100 foot level but below 850, the vein spread with only small ore bodies and further prospect work was terminated in 1887.

THE MEXICAN CANTINA

A small rustic building in Deep Gulch was a hang-out for the Mexican miners who consistently speculated their meager wages in games of monte and poker. The company restricted beverages to beer and wine. The absence of the professional gambler created a situation of friendly neighbors speculating against each other with their hard-earned wages.

congregated at an early hour in the plaza, where the ceremony began with a devotional service. The gathering then formed a procession and proceeded over the paths to the mines and other places of labor. During the course of the journey, each worker selected his site and placed a cross in the ground. Some entered the mines and placed crosses in the areas of their labor. Wherever this was done, the location was identified by the name of a special saint. Following the placing of many crosses, the ceremony was terminated until the evening hours when everyone assembled for a pot-luck, music and folk dancing.

Within the depths of the great cinnabar deposits, another custom prevailed for the Mexican miners, Being a dedicated group to their religious beliefs, the spiritual relationship was objectively expressed with strong feelings of devotion. The dimly lighted caverns and tunnels that gouged out the rich ore of Mine Hill were always a hazardous venture that challenged the trespasser. As the Mexican miners moved to various locations of operation, it was a practice to take with them some spiritual feeling of security. Within the cavernous walls of their labore, a sizeable niche was hewn from the hard, red rock and a shrine was installed with dedication to the holy protectress of the mine. This tutelary saint was clad in a white costume with red slippers, head-dress and ornamental decoration. The face, which may have been the creation of a village artisan, was a symbolic interpretation related to the subject. The facial characteristics were strongly accentuated by the prominent detail of the eyes which contained colored beads. The shrine was illuminated by the flickering light of candles arranged at the base. The lighting was consistently maintained by the replacement of candles for the duration of the work in this specific area. After the installation was completed and the candles had been lit, a simple ceremony by the group dedicated the shrine to "Nuestra Senora de Guadalupe." Thereafter, it was the practice of each worker as he entered the chamber of labor to prostrate himself at the base of the shrine in veneration and to supplicate the guardianship of the saint from prevailing danger.

Similar to other foreign groups who left their native soil to seek opportunity and a new way of life, the inhabitants of Spanishtown brought with them a feeling of nationalistic pride. Great events, that had considerable impact on the people and had influenced to a great extent the destiny of their country, became items of great concern and devout observance. In Spanishtown, there were two occasions which were observed with great exuberance that clearly expressed their nationalistic inheritance. The first was the "Cinco de Mayo," which commemorated the uprising and consolidated rebellion against the French ruler of Mexico, Maximilian. The other event was the "Diez y seis de Septiembre," which was acknowledged as Mexico's Independence Day. Both occasions were fervently expressed with color and ceremony. The Cinnabar Hills reverberated from the activity of expression as was typical of the Spanishtown inhabitants.

The final community function of the year took place on Christmas Eve or Noche Buena. This solemn ritual was called "Las Posadas" (the inn) which originally began on December 16 and continued until the 24th. The affair in Spanishtown was relegated to one evening. This rather primitive biblical translation, enacted by the combined gathering of the village, was a dramatization of the arduous journey experienced by Joseph and Mary and the seeking of shelter in the town of Bethlehem.

In preparation for the ritual, a large tray was constructed upon which would be placed the created images of Joseph and Mary. Early in the evening, everyone gathered at a specified location from which a procession would be conducted through the village. As the bearers of the tray, lighted by candles, led the way, the followers with lanterns and candles sang Christmas songs. They stopped at many houses for the purpose of requesting shelter but according to the planned procedure they would be rejected. After a short duration of this experience they arrived at the home, previously chosen, where they were accepted. There was the singing of songs and the recitation of a Litany. From the ceiling of the room was suspended by rope and pulley, a hollow clay object in the form of a rooster, which was called the Pinata. This term applied to any designed receptacle that was filled with "goodies." The occasion now changed from the solemn to one of humor and pleasure. The activity which now took place was prepared strictly for the children. After observing the dangling object, briefly decorated with colored feathers, they were led from the room. A child was selected, prepared with a blindfold, and given a stout stick about three feet in length. The child would attempt to strike the Pinata which was manipulated up and down by the rope. After several of the children had experienced no success, the Pinata was left in its original position for the final contestant who easily made the strike that released the contents. With the breaking of the stored "goodies," the children were rewarded for their efforts.

As the church bell in the plaza tolled the midnight hour, a small cradle was brought into the selected

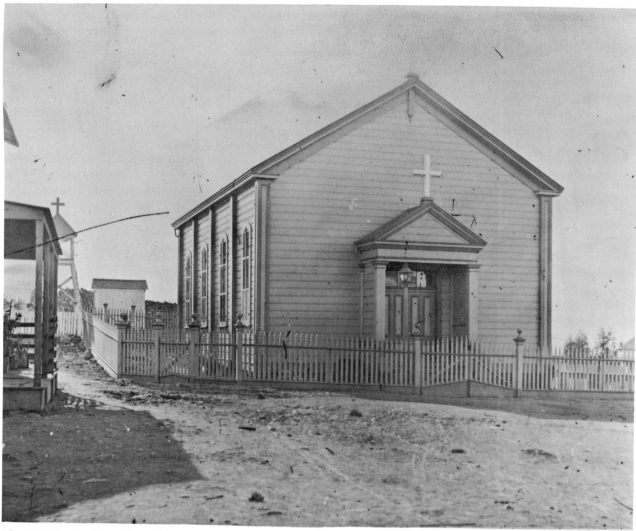

The Catholic church was located on the flat area of the hill overlooking Deep Gulch and the scattered dwellings of Spanishtown. This wood constructed edifice was built about 1885 by community contributions and company aid in money and supplies. It expressed simplicity in design with stained glass windows, an organ and belfry. The bell tower was separate from the building and at an early hour each Sunday morning the vibrant tone of the bell could be heard throughout the Cinnabar Hills.

home and a representation of the Christ child was placed in it. The group engaged in a short prayer and closed the ceremony with the singing of Ave Maria. The group then dispersed to their homes where the children were put to bed. The adults then continued on by lantern light to their little church on the ridge for the Mias de Gallo (mass of the rooster) or Midnight Mass.

One of the staple foods in the diet of the Mexican miners was beans. It was long a tradition in their native Mexico that the preparation of beans was a part of everyday living, and they were served in some form or other. Beans prepared with carne (meat) not only was a sufficient meal, but beans, also, were served as a side dish with Tacos, Tostados, Tamales, and Enchiladas. The cooking of beans for the Mexican housewife was a continuous chore in which was placed much per-

sonal pride. Using the red or pink bean, dependent upon the availability of either, she started the cooking with the beans in warm water. The beans were never soaked previous to cooking, unless, they were considered quite old. The cooking process was one of simmering rather than boiling, with frequent stirring. The cooking continued until the bean began to break open. At this time, a moderate amount of salt was added which generally occurs during the last half of the cooking process. Most of the housewives in Spanishtown used the small pink bean which seemed to be the most available.

Along with the consistent offering of beans, the Mexican housewife had other specialties such as the tortillas, the tostados, the tamales and the enchiladas. The tortillas could be considered the national bread and served as a basic part of many dishes. The early Indians of Mexico are credited with the

A sectional view of Spanishtown with great mounds of tailings excavated from Mine Hill.

invention of the tortillas long before the arrival of the Spaniards. The primitive prepared his corn on a stone metate, and it was made into thin flat cakes by patting with the hand. Related very closely, as to ingredients, was the tamale, which became adopted by the natives after it was introduced into Mexico by the conquistadores from Spain.

Throughout the years of habitation, Deep Gulch, with its steep sloped walls of red rock, native shrubs and scattered oak trees, contained within its perimeter, an assemblage of people who enjoyed their environment. Their confinement within the village was satisfying to their way of life. They worked, played and contributed their customs which gave color and life to the bizarre solitude of the Cinnabar Hills.

They were not isolated from the everyday commodities as most of their needs were available within their village. The religious service, to which they were fervent adherents, was adminis-

tered by a visiting priest from the Santa Clara Mission. Each Sunday, the populace walked to the high ridge where stood the small Roman Catholic church with its stained glass windows, belfry and organ. In the vicinity of the plaza, was a small rustic schoolhouse where one teacher taught about twenty-five children through grades one to four. After four grades the children completed their program in Englishtown. The idle hours of the miners were catered to by a cantina which offered card playing and billiards with a service of beer and wine. Other establishments that tended to make the settlement self-sufficient were a bakery, a tamale house, fruit store, barbershop, second-hand store and a shoemaker. Also, for the single men, a company boarding house was available.

When mining production could no longer warrant its existence, Spanishtown witnessed the gradual departure of its population and by the turn of the century, the little rustic houses became vacated, to survive their remaining years in the peace and quietude of nature's playful tactics.

DEEP GULCH

This scene gives an excellent perspective of the operations near Spanishtown. Above the massive dumps is the big Planilla and the scattered houses of the Mexican miners. At the left is the incline which descends to a winding track leading to the furnace yard. This scene was taken about 1887.

Open cut operations on the slopes of Deep Gulch, near Spanishtown.

THE MALACATE

This was a Spanish term for windlass or hoist. This scene shows the mule-powered hoist at work on one of the last prospect shafts to be sunk by the Quicksilver Mining Company. Photo in 1890's.

GIBSON GIRLS VISIT THE MINERS

This group picture was taken at the mine office on the Hill in 1897. The girls in bow ties and blouses, are dressed in attire that was characteristic of the pen and ink illustrations at the turn of the century by Charles Dana Gibson. L to R – Robert Bulmore, General Agent; Richard Harry, Foreman; Charles Derby, Superintendent; Ellard Carson, Surveyor; Thomas Derby, Proprietor, Almaden Store; Helen Mahany, Hattie Carson, Mrs. Charles Derby. Sated – David Mahany, President, Quicksilver Mining Company and David Mahany, Jr. Mr. Bulmore took this picture by remote cord as shown on the left.

On the Hill, all of the social functions and entertainment were held in the Helping Hand Club. In the center of this young group of entertainers is Charles O'Brion, prominent employee of the mining company.

CHAPTER 6

ENGLISHTOWN

A Unique Settlement of White Cottages and Colorful Gardens

High on the ridge, that was part of the hills called Cuchilla de la Mina, was situated the settlement of Englishtown. This was the termination of a winding, dirt road that followed the contour of the hill from the village of Hacienda. The title was derived from the assemblage of the inhabitants, most of whom had come from England and particularly, Cornwall. The two mile journey up a moderate grade was concluded at the center of the village which was composed of the general store, the mine office and other company buildings. As a mining camp, it was unique in its respective setting and the people enjoyed the amenities conducive to express a community of congenial habitation. Overlooking a panorama in all directions, the people enjoyed an environment and pattern of living unlike the typical mining camp. The healthful, moral and recreational demeanor of the people were greatly influenced by the initiative and attitude of the mining company. Saloons, dance halls and other features, that characterized the western mining camp, were inconspicuous. Mary Halleck Foote gives this account of the settlement in 1876:

"There is no undue propriety about the mining camps on the "Hill." Their domestic life has the most unrestrained frankness of expression and their charms are certainly not obtrusive. The Mexicans have the gift of harmoniousness; they seem always to fit their surroundings, and their dingy little camp has made itself at home on the barren hills, over which it is scattered; but the charm of the Cornish camp lies partly in vivid incongruity between its small, clamorous activities, and the repose of the vast, silent nature about it. . . . Many trees in the camp, standing at the meeting of the ways, bear upon their trunks certain excrescences in the shape of oblong boxes. To the New England mind they would at once suggest the daily paper; but the Cornish sustain life on something more substantial than 'bread and newspapers.' The meat-wagon, on its morning rounds, leaves Tyrrel his lego'mutton, Tregoning his soup-bone, and Trengove

his two-bit's worth of steak, in the boxes bearing these names respectively. Such is the honesty of the Cornish camp, that trees bearing soup-bones, leg-of-mutton and steaks, are never plucked of their fruit, save by the rightful owners."

Life on the Hill experienced a variety of weather. Under favorable conditions, one could see the glimmering waters of San Francisco Bay. The seasons were expressed in a natural cycle of change. The summer months were warm and the hills of sparse vegetation were covered with dry grass. During this time the water supply was dependent upon the storage tanks that were placed at strategic points about the area. During the winter season, occasional wind and rain storms played havoc, causing erosion and slides that disrupted the foot paths and roads.

At the time of the Gold Rush to California, mining activity in Cornwall was on the threshold of decline. The production of the copper mines was falling and competition from the mines in Chile, Australia and the United States caused many of the smaller mines to stop operating. Many of the younger miners, who had acquired the skills and techniques in deep, hard rock mining, realized a dismal outlook in the only work they had learned to do. As early as 1850, news had reached Europe of the fabulous gold fields in California. Many of the unemployed miners, who were free to move, began leaving for distant lands.

Conditions in Cornwall continued from bad to worse and the recovery of the mines to their former status appeared more remote. By 1866, the economic situation throughout England had reached a point of disastrous proportions. Business of all types was at its lowest activity. The copper mines were becoming fewer in operation and unemployment had increased to thousands of families subsisting on the lowest of rations. During the year 1866, an estimated 5,000 miners left Cornwall and in 1867, adverse conditions had multiplied and emigration continued. The once prosperous copper mines continued in decline and by the

THE NEW ALMADEN BENEFICIAL ASSOCIATION

This group composed a mutual and community service organization in Englishtown. The assemblage has congregated with "bowler" and badges for this picture taken in 1895.

Cinnabar Lodge 199, Knights of Pythias.

64

The Mountain Echo Band was organized on the Hill in 1890 made up of members from Spanishtown and Englishtown.

SONS OF ST. GEORGE

Members of the General Gordon Lodge, No. 286, Sons of St. George, assembled for this special picture taken in 1880. Most of the members are from the rugged shores of Cornwall. Church and lodge activity played an important part in this mining community of New Almaden.

They came from Cornwall and spread themselves from the copper mines in Michigan to the mining camps throughout the West. They were bred and dedicated to their trade in their native land. Their physical attributes for this rigorous work and the natural aptitude for hard-rock mining, placed them amongst the best in efficiency and endurance. This picture portrays the typical "Cousin Jack" with his candle holder and probably a pasty in the lunch pail.

When James B. Randol assumed the managerial duties at New Almaden, he immediately proceeded to obtain Cornish miners and during his regime a sizeable group became established on the Hill and with the preponderance of one nationality, it was given the title of Englishtown. The Cornish were reserved, friendly but somewhat clannish, similar to other nationalties who arrived on a strange soil to establish a new life. They were intelligent but in many cases illiterate, having started their life in the mines at an early age, with little formal schooling. Their physical make-up, stamina and natural aptitude, placed them at the top of the list among the ranks of miners. They were a dedicated group to the dangers and problems encountered in the depths of hard rock mining.

Once the Cornishman became established, he settled down to his work and place of living with a feeling of permanency. The impoverished conditions of his native land had created a frugal and conservative nature and the accumulation of savings became an important item in the Cornishman's mode of living. When he reached a certain stage of financial stability, he deemed it an obligation to send passage money to a relative. Such an event was common practice in most of the mining camps and the continuous arrival of kin soon created a large representation wherever mining was important. Jack, being a common name among the Cornish, it was not long until each new arrival from Cornwall was referred to as "Cousin Jack." Thereafter, as the Cornish became a prominent group of every mining town, the title "Cousin Jack" became the customary manner of reference, which they graciously accepted.

Englishtown was a neighborly settlement and congenial relations prevailed in this very compact and secluded community. There was a touch of the Victorian in the manner of dress and the decor of their homes. The integration of individuals, from various localities into this isolated environment, developed a mutual dependency in the social life, sickness and other situations of adversity. The one thing they shared in common, was their experiences and association with life of the mining camp. Although

turn of the century its economic importance in Cornwall had diminished.

For the clannish, conservative Cornishman, the departure from his native soil, heritage and traditions, to venture into a new land, was a momentous decision of speculation. However, the economic collapse and the prevailing conditions of adversity, offered no alternative. For most of the young unemployed that could finance passage, there was a continuous embarkation to the publicized mines of South Africa, Australia, South America and the United States. The greater majority would find in their new environment an opportunity for their skills and they would establish permanent roots, never to return to their native land. In the copper fields of Michigan and Butte, Montana, through the Mother Lode to Virginia City, Nevada and into the mining camps of the West, wherever they beckoned, came the Cornish miner.

they were self contained within boundaries of confinement, they strongly expressed an individuality and independence. This was the scene as observed by Arthur Lakes in the 1880's:

"The Hill:

The coach takes us by a winding path up a hill, the slopes covered with dark umbrageous oak trees, and at the summit is a little white church, reminding us of those we have seen similarly perched in the mountains of Switzerland. Around it is the village of neat white villas, each with its little garden, while white cottages peep out here and their on the steep hillsides from between dark oaks or the clustering vines. This is New Almaden, the prettiest neatest, cleanest, and most cheerful mining village we have ever visited. . . . Looking down from Mine Hill into the valley below we see by the character of the houses that there are two divisions in the little burg, one allotted to the Mexican miners, the other to the English speaking, which are mostly Cornish. Strolling about the town we are struck with the cleanliness and neatness of the place and with the absence of saloons and dance halls such as characterize and mar our western mining towns. There is an air of cheerfulness and sobriety among the people also and instead of pandemonium from saloons and dance halls, from nearly every house we hear music and singing in which the Cornish so much delight, while a band is playing in a substantial city clubroom hall and reading room and nearly every cottage has an organ or piano. The secret of this is that the town and all connected with it is in the hands and under the control of the company, and it is their aim, whilst keeping out things of immoral and hurtful nature, to supply their people with healthful recreation and elevating amusements. Such is New Almaden, a model mining camp."

The company built most of the homes which were scattered at random in the most practical sites on the sloping hills. The gable roof with a front porch was the standard pattern of construction. The houses varied in size from four to eight rooms and the rental charge was from $2 to $9 a month. There was a small number who built their own houses and paid a ground fee of 50c a month. The houses were available to employees only and the great majority remained in residence until the closing of the mines at the turn of the century. The houses were built of redwood with board and batten or grooved siding. Many of the cottages were enclosed with white picket fences and the display of colorful gardens and shrubbery was the custom for most occupants. Adding to the decor of the white cottage was the porch trellis with honeysuckle or climbing rose. Much of the garden foliage was furnished by the company who took a personal interest in the maintenance of each home.

These were the years before the advent of the gas range and electric light. Heating and cooking were dependent upon a good wood stove while lighting was by coal oil lamps and candles. Wood was plentiful in the near-by hills; it was transported to the furnaces and homes by six and eight horse teams. The miner could keep his home well supplied at a cost of $9 a cord. In 1881, the company installed a water system at a cost of $15,000 which served the mines and the homes. The water was piped from springs over a distance of three miles and distributed to 17 tanks. More than five miles of pipe was installed to facilitate the needs on the Hill. The water charge to each home was 50c a month. For some thirty years prior to this service, the people had to transport their own water or patronize a delivery service in which water was transported in small barrels attached to burros. Many of the Mexicans were engaged in this work.

The people were dependent for their supplies on the general merchandise store in Hacienda or Englishtown. The first merchant, to lease the company owned building, was a Mr. Brenham. The store soon after came into the hands of J. B. Randol. The flourishing production of quicksilver allowed little time or interest for the store business and Randol sold out his inventory to Thomas Derby, who later gave his clerk a partnership and the operation continued, until the closing of the mines, as Derby and Lowe, General Merchandise.

A wide range of items were sold at a reasonable profit plus transportation costs. While there was no strict rule requiring the local inhabitants to purchase all of their supplies at the local store, the practice appeared to be the customary procedure. It was considered a personal obligation to maintain a good will relationship with the local merchants. However, it has been stated by some of the old-timers that many of the housewives made occasional trips to San Jose in horse and buggy to shop for drygoods, clothing and household utensils. As they returned, they had secluded places near the end of the journey, where they would deposit their cargo. When dusk fell over the settlement, someone would descend to the spot, retrieve the articles and return unnoticed over a choice of footpaths.

The local merchants operated strictly on a cash and carry basis. Credit or charge accounts were not in vogue, but Thomas Derby did install a prac-

THE GENERAL MERCHANDISE STORE ON THE HILL

This brick constructed and company owned building was erected in 1864 as the supply center in the plaza of Englishtown. The building was leased to private merchants and in general offered a wide range of merchandise in food, clothing and utensils. Prices were slightly higher due to freighting cost. During the boom years the store was operated by Derby and Lowe.

THE BOLETO

The store on the Hill and in the Hacienda carried no credit accounts. As the company advanced no wages before pay-day, the store would issue small cardboard inscriptions called Boletos. A customer without funds, wishing to make a purchase was issued in different denominations, the desired amount. In these transactions, a record was sent to the company paymaster, who deducted the stated amount from the workers wages and sent the withdrawal to the store. The Boletos ranged in value from 5 cents to one dollar. There were occasions when workers wanting cash would draw Boletos and sell them at a discount.

THE MINING HEADQUARTERS ON THE HILL

The Quicksilver Mining Company conducted its mining operations from the Hill office located in the center of Englishtown. The Mexican on the burro is a Panadero or one who delivers bread to Hill-top customers. The wooden device over the burro's back is the bread container. This photo was taken in the early 1880's.

tice known as the Boleta system in which merchandise could be obtained on a pay-roll deduction plan. Patronage of this system was optional and was generally used only in emergency cases.

Within the store was a first class apothecary department which offered a quality of drugs and prescriptions comparable to other dispensers on the coast. The products and services were available at cost to all employees. The store, also, served as the distribution center for the mail that was brought up daily from the post office in Hacienda. As there was no delivery to the houses on the Hill, the people came to the store at a specified time each day where a clerk called out the names of the mail on hand.

While the company was partial and most receptive to the family unit, there were many who were unmarried. Accommodations were available to this group through a large company owned building which served the specific functions of a boarding house. The facilities were leased to experienced caterers who offered excellent menus at 75c a day. The second floor consisted of rooms to accommodate a limited number at a nominal charge. Verified peddlers were admitted to the property to supply the necessary needs of the boarding house with such items as vegetables, eggs, milk, poultry and meat.

The community center for recreational and social functions was the Helping Hand Hall. This building offered a large social hall with stage, a reading room and kitchen facilities. A small library contained 450 volumes related to history, science, fiction and juvenile literature. Weekly and daily newspapers included the S. F. Chronicle, S. F. Examiner, the San Jose Mercury, the San Jose Times, The Alta and the London Times.

All employees and their families who were contributors to the Miner's Fund were eligible to free use of the building. Most of the entertainment programs were initiated by local talent and, also, various functions were presented by the school children. Occasionally, special programs were brought in by musical and dramatic groups which gave the community a touch of professional talent. Most of the affairs were well mannered and conducted, and any event was a special occasion for the Hill people.

The following item, written by an unidentified Hill resident appeared in the San Jose Daily Mercury of June 15, 1886:

"The manager of the mine conceived the idea that by fitting up a nice, cozy and in every respect comfortable hall where all kinds of popular games could be indulged in and have attached to it a reading room and kitchen, that by so doing a great public want would be met that the social sphere would be enlarged in usefulness and our community become better natured and consequently more happy and contented. Our rules and regulations are very simple. Everybody and their families, if they have any, who pays $1 a month to the Miner's Fund are members and entitled to all privileges and can come to the hall when open, play games, read or take a cup of tea, coffee or chocolate at less than cost. No gambling or drinking of spiritous liquors is allowed. Smoking is allowed in the main hall but no games; no talking or smoking is allowed in the reading room. We have a library of 450 volumes, consisting of stories, biography, history, etc. Our magazines, weekly and daily newspapers comprises the best in the state or nation."

In contrast to the life in Spanishtown, the Englishtown inhabitants had no special days of recognition for celebrating. There was the occasion in 1887, when a special day was proclaimed to commemorate the 50th year of Queen Victoria on the throne of England. The company acknowledged the occasion by granting a holiday which was expressed in the simple festivities of a potluck picnic, with singing and instrumental renditions.

For most of the people, the social activities were closely associated with their church. There was the annual pot-luck picnic sponsored by the Methodist church. At the rear of the church under spreading oak trees, decorated tables accommodated a large gathering which took place in the afternoon and continued into the evening. This late afternoon affair was called a Tea Treat and was similar to that which had originated in England. Such an occasion brought out a variety of traditional recipes. During the evening hours the group assembled in the church basement for the climax which consisted of entertainment offered by individual members.

The Christmas season was always joyfully expressed by the total community. On Christmas eve, the Helping Hand Hall was decorated with a large tree trimmed with strung popcorn, paper chains and lighted candles. As there were no Christmas trees in the home, this community function served all the participants in expressing the spirit of the yuletide. The families assembled at an early hour in the evening loaded with packages of all sizes which were placed on the stage around the tree. Santa Claus, costumed in the traditional manner, assumed the lengthy task of distributing the great quantity of gifts by calling the names inscribed.

THE BOARDING HOUSE

This is one of several company owned boarding houses that accommodated the unmarried employees in the settlement. These buildings were leased for operation to private parties. As many as 100 boarders were served at the going rate of 75c a person. A limited number of rooms were available at a nominal charge.

Throughout the mining areas of the old West, the burro was a conspicuous member of most camps. Many of the younger generation in Englishtown and Spanishtown, were proud owners of this sturdy little animal. This line-up of riders was taken in front of the company boarding house in the late 1880's.

Following the distribution of gifts, the affair was closed by group participation in the singing of well-known carols.

On Christmas day, the residents assembled at the store where an organized group offered a variety of Christmas music. Much of the program was offered by the school children under the direction of their teacher. In the evening, the festivities were brought to a close by a group of young carolers who, with their leader, traveled the winding paths from house to house, finding their course by lantern light. Each home responded to this gesture by having available for their pleasure the customary homemade candy, cookies, saffron cake and tea.

Probably the focal point for attention and admiration on the Hill was the modest, well-built schoolhouse. The community showed considerable interest in the school and its function of providing education to the young generation. As many of the miners had received little formal education in their younger years, they were highly enthusiastic in seeing that their children received an adequate introduction to the school curriculum. Owing to the permanency for the great majority employed, the school population and attendance varied very little during some forty years of operation. Over these many years, most of the children received their total education, which consisted of grades one through eight. For some of the graduates who had the interest and aptitude, J. B. Randol sponsored a technical program to further prepare the student who was about to become a worker. The boys were offered vocational training in carpentry, blacksmithing and general mechanics. The girls were given classes in cooking and sewing.

The Hill school employed four teachers, one of whom was the principal. The curriculum was prescribed by the county board of education. Upon the completion of eight grades, the students were awarded diplomas by successfully qualifying in the examination issued by the county board. The school enrollment was at its greatest during the 1880's. In 1886, 253 students were in attendance. G. E. Lighthall was the teaching principal and also, served the community as Justice of the Peace. Adequate books and supplies were furnished to the students who were responsible for loss or damage. For those individuals who were school teachers in the average mining camp, the situation was one of many challenging circumstances. Many, who accepted such assignments, were the more rugged and adventuresome type, and their manner of strictness was necessary to survive. The control of the classroom was a strict disciplinary procedure

and it was not uncommon to observe the ever conspicuous strap to better impress their authority.

During the winter season, strong winds swept out of the gullies and ravines creating considerable impact as it passed the structures of Englishtown. The schoolhouse, being in open exposure to the elements of nature, was subjected to the cold and draughtiness of the environment. Within the classrooms, ventilation was reduced to a minimum and from the early hour in the morning, the pot-bellied stove glowed continuously. During the dark days, large kerosene lanterns and lamps provided the necessary light. This was the time of the year when the classroom atmosphere assumed an air of medication amid the prevalence of many odors. The students were prepared at home with various rub-on concoctions as preventative to colds. Also, many students would arrive day after day with the common and popular asafetida bags that hung under their garment from a string around the neck. This traditional practice was considered a sure preventative against colds, sore throat, diphtheria and other common ailments.

Life on the Hill for the younger generation was chiefly relegated to school work and chores. Recreational facilities for leisure time activity were of little consequence. Parental discipline established a certain routine of chores, in and about the home. A ten hour shift for the miner left little time for the maintenance of the home activities and for this reason, the children were indoctrinated at an early age in assuming certain responsibilities. The girls, in their early years, became proficient in all areas of cooking, sewing and general household routine. The boys took care of the wood supply and, in some cases, milked a cow, raised rabbits or took care of the chickens. Some of the boys were fortunate in having occasional chores for pay offered by the company. Life was not dull for the young generation of the Cinnabar Hills.

During the years of operation by the Quicksilver Mining Company, law and order were of little consequence in New Almaden. As the people in residence were all employees of the company, their livelihood and employment required strict adherence to specific rules and regulations. Individuals who became trouble-makers or undesirable to the interests of the company were discharged and notified to leave the premises. With this order of discipline, the environment in general maintained a favorable atmosphere of order.

However, as New Almaden became an established township, a deputized officer was appointed as constable to handle any necessary situations of misdemeanor. Arrest cases requiring legal judge-

PICTURE DAY AT THE HILL-TOP SCHOOL

This photo was taken about 1888 by Dr. Winn, medic for the mining company. The schoolhouse was situated on an incline facing the plaza. For over forty years, it served the educational needs of the young population.

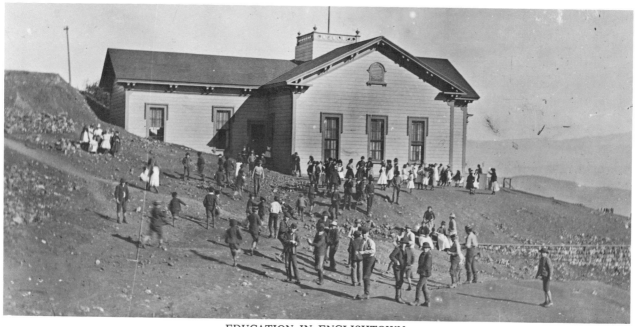

EDUCATION IN ENGLISHTOWN

An Elementary School District was established on the Hill in 1864 and this modest building was erected on a site overlooking the town center. This scene depicts the setting in its early years quite soon after completion.
No ground work or landscaping has taken place.

A view of the schoolhouse following the completion of fencing and landscaping as it looked in the 1880's.

The schoolhouse in the 1870's.

ment were presented to a Justice of Peace who conducted his business in the Hacienda. From the year 1856 until the population left the settlements. New Almaden life carried on with a fair degree of law and order. A small jail accommodated overnight guests but except for sporadic or spontaneous occasions where trouble, in many cases, was instigated by outsiders, the duties of the law were quite simple and congenial.

One of the special events that took place in many of the mining camps of the old West was hand drilling contests. Such was the case in Englishtown at least once a year when the contestants met to match their skill and dexterity. The miner who specialized in drilling hard rock walls for the purpose of placing a charge of explosives, became very adept in the techniques of the operation. As most of the work in the mines was on a contract basis, the miner became highly proficient in the manipulation of tools by the daily challenge of increasing speed.

A drilling contest was a competitive activity for individual honors, an award and, also, served as a form of entertainment for an audience. The contest was conducted under the supervision of several judges in accordance with specific rules. On the day of the event, a large representation assembled in the picnic area to stimulate the efforts of their favorite candidate. Audience participation involved a certain amount of wagering and many silver dollars would change hands during the course of the event. A large block of granite or stone of similar hardness, having two flat surfaces on top and bottom, was brought in for the occasion. Most of the contestants worked as two men teams who had worked together deep in the hard rock formations. The contest was restricted to a fifteen minute time limit in which fifteen drills would be used with a change of drill every minute. An experienced team would function in a pattern of rhythmic unison. One miner held the drill which he turned slightly with each strike of his partner's eight pound double jack hammer. It was generally a practice of most teams to alternate positions on several occasions during this fifteen minute ordeal. Many experienced contestants could make this change without missing a strike or breaking the rhythm of the procedure. At the termination of the time limit, each hole was measured and recorded. The grand prize was awarded on a descending scale which might range from one to five, according to the number of contestants competing.

There was an opportunity for the individual to display his skill in single hand drilling. The same rules governed the contest. A short single jack hammer, weighing about four pounds, with three-quarter inch drills were used. This type of competition was a test of stamina as well as the skillful use of the tools.

Each nationality was characterized in a new community by their customs, temperament, mode of living and special recipes for food. With the great number of Cornish in Englishtown, several traditional food items were an important part of the household menu. The most prominent invention of the Cornish housewife was the Pasty, which had its origin in the early mining days of Cornwall. These oven baked specialties constituted a substantial meal, conveniently designed to be carried into the mines, to the sea for fishing, to a picnic or for the family table. In many families, the making of the pasty was generally dependent upon the desires of the individual and each was made according to his own fancy. When the pasty varied in contents, it was customary to place an initial for means of identification on the crust when put in the oven. While the pasty might vary in recipe, the basic ingredients were lean beefsteak, potatoes and a touch of onion, carrot, or turnip. According to the viewpoints of an expert, the making of a pasty was an art. These carefully designed creations enclosed within a semi-circle, oven browned pastry, gained favor with many people in the mining camps. A Cornish pasty formed a meal which could be eaten underground. Wrapped in a clean cloth, it required neither knife, fork, nor plate.

The defense had probably found the story told in "Q's" first number of the "Cornish Magazine," June, 1898:

"A miner whose mother had daily made for him a pasty for 'crouse,' married a cook who had been in the service of a wealthy family. On going to the mine after his marriage he took with him a pasty of his wife's making. 'How did you like the pasty?' asked she when he returned in the evening. 'Ed wadn no good,' replied the miner. 'Time I got down to the fefty fathom 'ee was scat to lembs. The wans mother made wudden brack of 'em falled to the bottom. They was pasties.'"

Some lines in the same magazine, by the late Morton Nance, describes the value of the pasty:
"When the Tinner to Bal takes a touchpipe for crowse
He cannot have Hot-meat sent for his house:
Yet hath no stomach for victuals cold,
So a Pasty he takes in a napkin rolled;
And though he leaves it for half the day,
Within his Hogan Bab warm 'twill stay."

The typical recipe and customary procedure brought from Cornwall was as follows:

High above the roof tops of the miner's cottages in Englishtown, stood the Methodist Episcopal church that faced a northern panorama of the Santa Clara Valley. This picture shows the last of three structures that served the religious and social life of the community. The first church was built about 1871, but during its first winter a severe rain and wind storm wrecked the structure beyond repair. A new church was built in 1884, but in a space of a few years, fire razed it to the ground. The third and final structure was erected at a cost of $3400 which was financed by contributions from the community and the mining company. This structure remained standing in ghostly silence for many years following the closing of the mines and the departure of the inhabitants.

(Courtesy Suzie Gilman Collection)

The general store and the Methodist church.

Looking north over Englishtown can be seen the schoolhouse upper left and the large boarding house in the foreground. This was the scene as it looked about 1885.

This is Englishtown as it appeared about 1890 looking to the Mt. Hamilton range in the background. Surrounding the general store in the center are the miner's homes and company buildings. At the bottom, (left to right) is the dormitory, Helping Hand Club and boarding house.

1 lb. plain flour
½ teaspoon salt
4 oz. lard
4 oz. shredded suet
½ lb. lean beefsteak
4 oz. potatoes
1 onion

Sift the flour and add salt. Add lard and suet to flour and mix to a dough using cold water. After the dough has been well kneaded, roll out fairly thin. Cut circular shapes from the rolled dough six to eight inches in diameter. Medium sized plates will simplify the cutting of circles.

The preparation and quantity of the ingredients depended upon the number to be made. Only with experience can the amount of contents be determined to meet the required servings. The addition of chopped onion or carrot is used with discretion according to the likes or dislikes of the individual.

Good quality, lean beefsteak is prepared by cutting into cubes averaging one inch in size. Slice potatoes and cut to about the same size as the meat. Place meat and potatoes in a bowl and mix together. Add to this any optional selections, if desired. The addition of the onion is used sparingly or can be omitted. Place a compact portion on half of each round of piecrust and add a touch of salt and pepper. The open part is brought over the contents and the edges are shaped together by pressing one edge over the other. A neat closing of the pasty can be made by applying the flat of a fork along the edge. Several small slits are made on top to allow vents for the steaming contents. Arrange in flat pan and place in oven at high temperature for twenty minutes after which reduce heat to moderate and bake for about one hour. During the last fifteen minutes, add a little water to the vent openings to moisten the contents.

If the basic details of preparation and baking have been given proper consideration, the pasty comes from the oven with a light brown crust that has held firmly together and is a true picture of Cornish delight.

The housewife, also, had a traditional specialty which was called Cornish Cream. In their native land, which had a long season of verdant grazing, the milk production contained a very high percentage of butter fat. The abundance of rich cream and the frugality of the people probably brought about this ingenious method of conservation. Some of the families in New Almaden had adequate facilities to have a cow. Following the winter season, the open hills adjacent to the settlements were lush with fresh, native grass. It was not an uncommon practice for many of the boys, before the school bell rang in the morning, to take their cow to a green pasture where it would be staked out until late afternoon.

The making of Cornish Cream was a simple chore and its application to cereal, fruit preserves, pie and other dessert dishes, made a very palatable treat. One prime essential necessary for successful results was sufficient cream in the milk supply. The customary procedure was followed by most makers of this product.

A selected amount of milk is placed in a bowl or container and allowed to remain all day in a cool place. The top half is then skimmed off and put in a bowl which is in simmering water or in the top half of a double saucepan. After this gradual heating, the surface proceeds to develop a thick skin. After about four hours, the milk is taken off the heat, using caution not to disturb the crusty covering. The container is now put in a cool place until the contents have thoroughly cooled. After a thorough cooling, the surface crust is skimmed off and the final product is finished.

Another delectable treat with the Englishtown inhabitants was saffron cake or bun. It became a custom for the Cornish people to express the advent of Good Friday with a display of saffron buns or a large flat cake, either, of which, was glazed over the top with a mixture of powdered sugar and milk. "As yellow as saffron" was a common saying because of the high intensity and brilliance of this particular color. The saffron plant is a species of crocus that produces a purplish flower. The stigmas of this plant were collected and dried and became much in demand for the dyeing of cloth and the flavoring of food. In most situations, the purchase of saffron was limited to the drug store.

The typical Cornish recipe and procedure for making saffron cake was as follows:

8 cups of sifted standard flour
1½ cups of shortening
¼ teaspoon of saffron color
2 teaspoons of grated nutmeg
1½ teaspoons of salt
½ cup of dried currants
½ cup of shredded candied fruit rind
½ cup of sugar
1 cup of scalded milk
1½ cakes of compressed yeast, dissolved in ¼ cup of lukewarm water
1 teaspoon of sugar

Break or grate the yeast into a bowl and add ½ cup of lukewarm water containing one teaspoon of sugar. Place in a warm place until the mixture is light and spongy. This generally takes about fifteen minutes.

Mix the shortening, sugar, salt and nutmeg in a large bowl and add the hot, scalded milk. Stir the contents until well mixed. Add the yeast mixture, saffron coloring and the flour. Mix well and add shredded fruit rind. The dough should now be thoroughly kneaded and placed in a greased bowl, giving the top surface a light brushing of shortening. Cover dough and place in a warm place for rising until the contents have doubled in size (about two hours).

When the dough has risen sufficiently, it can be shaped into buns or placed in greased cake containers. The top is then brushed with an egg-milk mixture and left for another rising (about one hour).

When the final rising has been completed, place in hot oven at a heat of 400°F. for fifteen minutes and then reduce to a moderate 350°. The size and condition of the dough will determine the final baking time but it will require a minimum of twenty minutes or more.

The everyday life in the miner's cottage was a routine that varied little, week in and week out. The needs and desires for most of the families conformed to a pattern of minimum cost which constituted a very plain, but substantial type of living. A great majority of the inhabitants on the Hill were in their younger years. They were well adjusted to the rigors of mining camp life which demanded a certain dedication of spirit and energy. While living was a simple formula, the hours of activity were long and the tasks many in this small and self-contained society.

The housewives of Englishtown were a busy group because of their isolation from any urban conveniences. She was obliged to substitute her talents and energy in acquiring the basic commodities, which in the present day, are unlimited in the super-market. The raising of a family and meeting the obligations of the home, was a daily challenge of stamina, which they accepted with pride and devotion.

The first prerequisite was the knowledge and ability in the daily preparation of food. The standard cast-iron, wood burning stove was a permanent fixture in every kitchen and for the Almadeners, it was never replaced. Bread making was almost compulsory and the weekly production might range from five to twenty loaves, depending on the size of the family. This weekly chore might occur on several occasions. In all phases of baking, great

pride was taken, and baking and cooking recipes were freely exchanged amongst the neighbors. Every housewife had been well oriented to this work during her younger years and recipes and techniques were passed on from mother to daughter.

Probably, the most menial task of the housewife was the upkeep of clothing, bedding and curtains which made wash-day a laborious event by its primitive methods. Most of the items were boiled in a copper boiler on the stove and were rotated at intervals by a wooden stick. From here the wash goods were scrubbed on a corrugated, metal wash board in a tub of warm water and then were rinsed and hand wrung for drying. A large family wash was not always solely dependent upon the women and in many situations the young boys and girls served as helpers in this somewhat arduous chore. Following the drying of the wash, another task of no minimum proportions, was that of ironing. For this work, the heavy, cast-iron iron was used which required a specially made hand pad for handling. Several irons would be continuously heated on the stove. The older girls in the family learned the techniques of ironing and assumed the duties as part of their growing-up experiences.

Store-made clothes for girls and women particularly, were a rarity. Some skill in sewing was a necessity in every house. Because of the accumulation of flour sacks, this sturdy material served to supply nightgowns, underclothes and other items of use. In some cases unbleached muslin was in common use. For special occasions such as Easter, Christmas or school graduation, a trip would be made to the city where silks, satins and colorful printed material would be available. In the art of sewing, like that of cooking, most of the women were instructed in their early years to the various techniques of needlework. In all of their creations, great pride was taken in producing a flawless product. They were efficiently skilled in stitching, embroidering, crocheting and knitting. Also, in their leisure moments the women would contribute hours in piecing quilts. This was an item that each housewife was proud to have on display which reflected her sense of color and design. In the same area of endeavour, was the making of braided rugs which was conspicuous on most cottage floors. Old clothes and cloth remnants were always saved for this activity.

While there was milk delivery to the Hill from nearby ranches, many of the miners had a cow. Another practical activity of the housewife which was at least a weekly chore was the making of butter. Using a wooden paddle and a bowl she produced and maintained the butter supply for the

A sectional view of Englishtown. These little cottages with colorful gardens and white picket fencing were characteristic of the hill-top settlement.

A SECTIONAL VIEW OF ENGLISHTOWN

This view shows the north slope of Englishtown taken from the schoolyard about 1885. In the foreground is the mining office. On the distant hill is the Methodist church.

A MINER'S COTTAGE

Clinging to the steep incline, this vine-covered cottage is typical of the many that were accessible only by foot-path. Mexicans, with their mules heavily laden with wood are making deliveries.

A sectional view of Englishtown.

family. During the fruit and berry season, prices permitting, many of the cottage shelves were stacked with jellies and preserves.

As there was a certain feeling of permanency on the Hill, the women maintained their rented cottages with a pride of ownership. The bare, wooden floor of the kitchen was hand scrubbed at regular intervals. Lace curtains hung on the windows and the walls were papered in colorful patterns. The large kitchen table was always covered with plain or patterned oil cloth. The bedrooms had full floor carpeting, high board beds and a commode. The special room, where was displayed the most treasured belongings, was the parlor. Here, assembled the guests for visiting or dining. The table in the center might be covered with a lace tablecloth and always in plain view was the large, family bible with flowery designed pages for births, marriages and deaths. If the family were well established, a small organ or piano would be the center of attention. As kerosene was the main fuel for light, an ornate suspension lamp with a hand painted, glass shade would hang from the ceiling. There was always a rocking chair or two that served in the house and on the front porch. As practically all of the homes were built with a covered porch, every family made much use of this convenience during the long summer evenings.

Throughout the village, the housewives maintained a competitive spirit in the display and maintenance of their home. Their skill and proficiency were rewarded by the satisfaction they derived from the charm and coziness of their home whereby, they enjoyed the amenities of everyday living quite uncommon to mining camp standards.

For the miner, his vine-covered cottage was his castle. As he returned from the mines with skin blackened by powder smoke or caked with the red dust from hours of perspiration, his spirits heightened when he entered his white, picket gate. On the back porch would be a small tub on a bench and warm water on the stove. Here he would remove the grime of the day's labor and change clothes for the remaining hours of leisure. When not relaxing on the front porch or having a confab at the store front, he might be attending a lodge meeting or scanning the reading materials at the Helping Hand Club. Life on the Hill was never dull and the miner had many advantages to satisfy his interests. For these people, who were devout Almadeners, they felt completely secure, even though the rumors were becoming evident that the future days of the Cinnabar Hills did not appear promising.

The closing days, of the great cinnabar deposits, gradually arrived with the scarcity of good ore and the people in Englishtown began looking over the ridge to new horizons. Mine Hill had been burrowed very methodically with more than one hundred miles of excavation. For the many miners who spent years in the New Almaden mines, the signs of depletion became more obvious as the tonnage of ore continuously decreased. Further development to greater depths was not within the budget or optimism of the company to carry on exploratory work.

The drastic means of reducing operational costs, brought about a consistent decrease in employment. Many of the miners, seeing the imminent outcome, began to move out; others, hoping for better days, held on to the end.

With the closing of operations by the Quicksilver Mining Company in 1912, Englishtown was all but deserted and only a few families held fast to the little cottage they found difficult to leave. However, the days were numbered for life on the Hill and it was not long until the hill-top village became silent by the absence of its population. The well-kept cottages and company buildings became a picture of solitude, awaiting the forces of nature and its relentless process of transformation. The propagation of nature's foliage became rampant as it took over the gardens and filled the areas once bared by footsteps.

In a few years, the colorful setting of Englishtown was a pathetic picture of weathered structures, deteriorated beyond repair. Curious visitors and the work of vandals, stripped away or destroyed anything of value. Without windows or doors and the walls bleached by the summer heat, the buildings crumbled away until they would completely disappear at the hands of man.

Today, Englishtown exists only in the thoughts of a small number of descendants, who were born on the Hill where they spent their early years. Nothing remains of substantial evidence that would indicate to the casual observer, that on the ridge was a little world of contented people who had maintained a congenial way of life for half a century. It is fortunate that the forgotten settlement has been retained through the medium of photography, for the Cornish miners and their families left little record of life on the Hill.

Small reading room in the Helping Hand Club where the employees in their leisure time had access to books, magazines and several newspapers.

DIA DE RAYA 1886

The employees of the mine received their wages at the close of each month. This was pay-day which the Mexicans called Dia de Raya. In this photo, the line is forming at the mine office in Englishtown to receive their pay in silver dollars. The Derby-Lowe general store is in the background.

(Winn Photo)

THE HELPING HAND CLUB ON THE HILL

This interior view of the large club room was taken in the 1880's. Every convenience for recreation and leisure time was offered to the employees by the company. James B. Randol initiated the construction of this building in 1886.

PAY DAY

This scene was taken in the 1870's and the occasion is another pay day at the Mine Office on the Hill.

THE SCHOOL ON THE HILL IN ENGLISHTOWN

These photos were taken in 1895 by John Tucker, commercial potographer from San Jose. The above picture is the lower grades. Miss Roberts is the teacher. The picture below is the upper grades. Mr. Chaplin is the teacher and principal.

A westward view of the schoolhouse and company buildings.

REMNANTS IN SOLITUDE

This photo, taken in 1927, shows the last years of the few remaining cottages and the crumbling walls of the old store. By 1912, the last inhabitants had departed, leaving the remnants of Englishtown to endure the elements of nature until the early 1930's, when the delapidated remains would be cleared away by a government project. Today, there still stands a section of the brick-walled store and the outer structure of the schoolhouse. The scene has reverted back to nature's design of growth and vegetation and only the contour of the hills serves as any means of identification.   *(Milton Lanyon Photo)*

The well-kept schoolhouse which was the pride of the Hill population had become but a ghostly remnant by 1927.

(Milton Lanyon Photo)

This setting was the center of activity for the people on the Hill during its half century of existence. Hidden in the towering pines is the schoolhouse. Behind the spreading oak is the general store. The Company Office at the right no longer awaits the miners on Dia de Raya. This picture taken in 1924, is a true expression of ghost town environment which lasted some twenty years.

(Courtesy Carl Nipper)

A scene in Englishtown showing the hose cart house with its fire bell still suspended. *(Courtesy Carl Nipper)*

THE TWILIGHT YEARS

This photograph, taken in 1924, was a scene of ghostly silence. The customers have long departed and the passing of time has left a remnant of what was once the supply center for the inhabitants. Above the store on the hill the old Methodist church is rapidly succumbing to the elements of nature and in a few years will be razed for its lumber.

(Courtesy John Gordon)

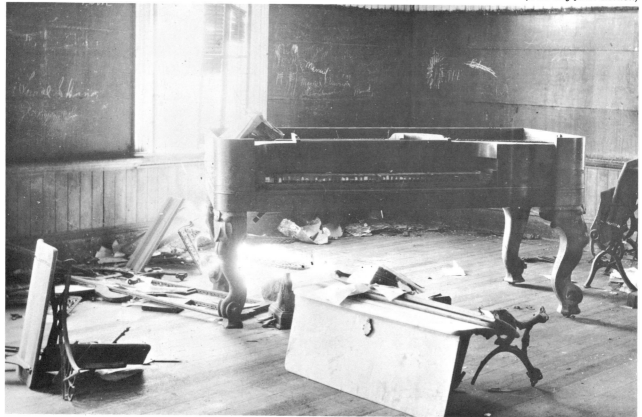

NO STUDENTS — NO SCHOOL

School days on the Hill are but a memory to the small group of graduates available today that recall the learning of the three R's in the mining camp of New Almaden. The old schoolhouse was deserted with its contents intact for the casual observer and wandering vandal that would occasionally make visits to the open premises. This classroom interior was photographed in 1924 and the last days are approaching rapidly for soon all of its contents will have disappeared.

(Courtesy Carl Nipper)

ALL WAS NOT WORK FOR THE RUGGED MINER

Relaxation from the rigors of hard rock mining was expressed in various ways by the people of Spanishtown and Englishtown. The Mexican population had many more occasions for celebrating than did their conservative neighbors. This photo taken about 1890, shows a group of Englishtown miners in preparation for a "Hard Time" party.

The New Almaden rifle club. Hunting and fishing was a popular activity in this mining area.

This is a scene of Hacienda taken about 1904. The dirt road thoroughfare leads through the village to the Reduction Works and over the hill to the mines. Mining activity has reached borrasca and soon many of the cottages will be vacated. Clearly visible in the scene are the store, mine office and reduction works which appear from the exhaust on the hill to be in operation.

(Watkins Photo)

CHAPTER 7

THE HACIENDA

The Gateway to the Mines

When the Barron, Forbes Company prepared for mining operations in 1847, they established their headquarters at the base of the hills on a tree-lined stream which was titled in Spanish, the Arroyo de las Alamitos (the Little River of Poplar Trees). The site was selected as the best suited for the building of the furnaces. The location was referred to as the Hacienda de Beneficio or Reduction Works. With the passing years and growth of the settlement, it commonly became known as the Hacienda. The location served as a terminal for the general operations of the mines and the reduction of ore at the furnaces.

In 1852, the mining operations were well under way and the area was identified as the Almaden Township. John Young and D. F. Winslow became the first Justices of the Peace. George Day and John Bohlman were appointed Constables. With the isolated community established in Deep Gulch, the primitive and disorganized conditions, made the need for law and order imperative. This secluded settlement was well known as a hide-out for desperados and their visits were frequent, after terrorizing activities throughout the county. The secluded hills offered seclusion from the law and many were sheltered by friends.

The Hacienda had little development until the change of ownership in 1863. With the acquisition of the mines by the Quicksilver Mining Company and their manager Samuel Butterworth, the settlement was immediately expanded with the building of wooden and adobe cottages. A general store was established to accommodate the population with necessary supplies. The little settlement soon became a cosmopolitan assemblage of many nationalities spread along the main thoroughfare. Running along the front of the homes was a man-made ditch called the Acequia, where water from the furnace yard coursed its way through the town and returned to the creek at the outskirts. It was a source of water for the cottage gardens and a playground for the children.

Life was quiet in the Hacienda as the greater percentage of the population lived on the Hill. The general environment was pleasant and its pic-turesque setting became a showplace at the gateway to the mines. In July, 1861, the first Post Office was opened with John Brodie as the first Postmaster. On the main street was the Bohlman livery stable which served as headquarters for stage service, in and out of the settlement. The Justice of the Peace and the Doctor shared a small cottage for conducting their business. For community gatherings, the Helping Hand Club served for many social functions. In the later years, the Spanishtown Cascarone Ball was held here with traveling musicians from San Jose providing the music. The heavy ore wagons, teamsters with their loads for the railway depot and vehicles transporting passengers, made the thoroughfare a continuous source of red dust that permeated the homes.

On special occasions, community dances were a big event either as a social affair or for the purpose of raising funds. The feature attraction that stimulated an enthusiastic response was the appearance of Tillie Brohaska and her eight piece orchestra from San Jose. Tillie, an accomplished musician, spent a full life of many years in musical entertainment and her talents were in great demand during the horse and buggy era. Her group of musicians performed on many occasions for the Almadeners at the Hacienda Hall. In her later years, following retirement from musical activity, Tillie Brohaska briefly stated by letter in 1946 of her engagements at New Almaden:

"Our visits, to the Hacienda to provide dance music, was always a diversion of interest. Anything could happen and it generally did. The people came out of the hills with much vigor and enthusiasm as an occasion of this type was considered an important event in their rather quiet life. It was not uncommon to find that pranksters had tampered with the piano or had placed articles under the strings to effect the playing.

There was one engagement which I remember very well. As we made preparations to get the affair under way, the piano was opened and a large rat jumped out and scampered about the floor creating havoc amongst a

91

THE HACIENDA SCHOOLHOUSE

This two-teacher school was located on a flat near the Casa Grande. The enrollment came chiefly from the Hacienda with a small number from nearby ranches. The total number of students averaged about 85. Photo about 1886.

Frank Bohlman and family at their cottage along the shaded Acequia. Mr. Bohlman was owner of the livery stable and contracted all of the transportation work for the mining company.

THE HELPING HAND CLUB IN HACIENDA

In 1886, mine manager, J. B. Randol initiated a move to construct a recreational hall in Hacienda. The building was constructed by Giles McDougal of San Jose. The lower floor offered a large assembly hall, game and card room and kitchen. The stage was arranged with incline from the front and was patterned after the Tivoli theater in San Francisco. The upper floor offered well furnished rooms for over-night visitors to the area. The building stood for many years following the closing of the mines and was dismantled in 1941.

The tree-shaded thoroughfare through the village of Hacienda was most conspicuous for its well kept cottages of which there were two types; the modest wood construction and the adobe brick and plaster. Each cottage expressed the individual touch of the occupant in color and decor. Colorful gardens and rambling vines attached to porch trellis work enclosed with white picket fences gave a charm quite unique for most mining districts. This scene is a typical dwelling as it looked before 1900.   *(Winn Photo)*

The general store in the Hacienda as it appeared in 1885.

The general store as it appeared in 1937.

crowded audience. After much commotion, the rat finally found exit through an open door. When things calmed down, we started again but I found that a section of keys would not play. The music was stopped for an inspection of the piano. A rat nest with two baby rats was discovered tucked in the keys. Intermission prevailed until I was certain that the old piano was free of trouble.

The dance got under way and everything was going fine until there was a raising of voices and a scuffle took place. We kept the music going until several shots raised the pitch of the commotion. At this moment, my bass fiddler dropped his instrument, jumped off the platform and disappeared through the milling crowd. We had another intermission until the situation was brought under control by the town constables with the help of some of the audience. When order was resumed it was found that one of the shots had passed through the bass fiddle. The music continued for the balance of the night without the fiddler. Selections for the waltz, polka, schottische and square dancing rounded out the program.

When the affair came to a conclusion, we were concerned as to what happened to the lost fiddler. Small groups with lanterns began an intensive search about the buildings and over the winding paths. It was not until the pre-dawn hours that the lost fiddler was found some distance away in a small picnic area. Shivering from the coolness of the early morning hours and emotionally affected from his narrow escape, he was returned none the worse for his experience.

While we performed many times later, I could never get the bass fiddler to return to the Almaden Hills. Yes, indeed, a trip to the mining village was always something to look forward to."

Leo Sullivan, who is retired from many years in orchestra and theater work, had this comment of his visits to New Almaden:

"Yes, I remember several trips to the Almaden country in the late 90's as a member of Tillie Brohaska's orchestra. I was just a "kid" in my "teens" when I started this type of work and it was quite an experience for a beginning musician. The journey to Almaden was twelve miles by horse and buggy. We would leave San Jose and arrive at the Hacienda in about two hours. In those days, six miles an hour in horse and buggy was considered a fair rate of travel.

The people were always gracious when we arrived and served us an excellent meal in the old hotel before the evening program started. Most of the affairs were well mannered and everyone made a big event of the occasion. Of course, there was always the unexpected, and disturbances and brawls were not uncommon. But then, this was still a part of the old West. When the program was over we were always treated to a big tamale feed and they sure knew how to make tamales in Almaden.

Well, that's a long time ago, but the trips to Almaden and especially returning home was something not easy to forget. I remember dark, rainy nights when you couldn't see the road. We didn't have much in the way of lights in those days. If it hadn't been for good horse sense, we never would have made it. Yes, Almaden was quite a place and the people sure liked an occasion of entertainment. As I look back, it was an experience worth remembering."

Mary Hallek Foote gave this impression of the Hacienda as she observed it in 1876:

"The charms of the Hacienda are of the obvious kind; a long shady street, following the ripples of a stream spoken of as the "Arroyo de las Alamitos," at one end the manager's house, with its double piazzas and easy hospitable breadth of front, a lonely background of mountains at the other, and vine covered cottages between. These agreeable objects can be well appreciated in a drive along the main street as in a year's residence there, it is very pretty; but as the "show" village of the mine, ever conscious of the manager's presence, the Hacienda wears an air of propriety and best behavior, fatal to its picturesqueness."

Upon entering the shaded promenade of the Hacienda leading to the great furnaces, stood the pretentious structure of classic symmetry, called the Casa Grande. This building originally was planned as a hotel, but the plan failed to materialize and instead, became the personal and official residence of the mine manager. During the years of occupancy, the different managers enjoyed a unique, rural splendor uncommon to mining environments. Nowhere, throughout the western mining areas was a touch of grandeur so simply and artistically expressed, as this stately building in New Almaden.

Through its doorway, came many visitors and dignitaries to visit, for personal or official reasons. Such prominent personalities as William Sharon, financier and builder of the Virginia and Truckee

PRETENTIOUS ELEGANCE IN THE CINNABAR HILLS

At the gateway to The Hacienda, stood the stately structure of massive stone walls, called Casa Grande. Planned originally as a hotel which did not materialize, this building, constructed by Francis Meyers in 1854, became the residence of the company managers. The five acre setting was formally landscaped with lawn, flower gardens and shrubs bordering the Alamitos creek from which water was diverted into a man-made lagoon. This sturdy structure still stands today and continues in service as Club Almaden. The building and grounds have been altered to offer recreational facilities to the public at a nominal charge.

THE LAGOON – 1892

Situated at the rear of the Casa Grande, was the man-made lagoon enclosed by colorful gardens. While this was a noticeable point of interest to the landscape, it also was available for boating and swimming. Today, the same location has been converted into a large swimming pool for public patronage.

Railroad into Virginia City, Nevada; James Fair, Comstock silver king; William Ralston, banker and builder of the Palace Hotel in San Francisco; Baron Rothschild, a strong manipulator in the world's quicksilver market; an emissary of the Emperor of China; Thomas Bell, a prominent figure in western finances and sales representative for New Almaden quicksilver; foreign consuls, politicians and many others of lesser prestige, gave the Casa Grande a touch of dignity quite exclusive in the world of mining camps. Through the years of turbulence, litigation and quicksilver production, this imposing structure maintained a certain prominence and appeared to be a welcoming shrine at the gateway to the Cinnabar Hills.

Situated upon a setting of five acres, landscaped with lawns, flower gardens, trees, shrubs and a family orchard, the life expectancy of the Casa Grande seemed indefinite. Many of the garden specimens were imported from foreign countries and the prolific blossom and foliage showed a natural affinity for the rugged environment. John McLaren, assistant superintendent of Golden Gate Park in San Francisco in 1886, visited the property many times and contributed his services as an adviser, in the further development of the landscaping. A short distance from the rear of the building, the Los Alamitos creek was diverted by pipes into a large man-made lake. Robert Scott, the designer and builder of the furnaces, created this colorful addition to the property. The water, averaging five feet in depth, was sufficient to accommodate many swimmers and small row boats. Within the lake, a large tree trunk proved too difficult to remove and was converted into a small island.

The Casa Grande was designed and built by Francis Meyers, under the direction of Captain Halleck, in 1854. This structure of stone and wood, was constructed on a rectangular foundation and consisted of two floors. Due to the inclined site, the rear of the structure was spacious enough for a large basement which contained a kitchen, storage room, a wood-burning furnace, servant's quarters and a large vault. The exterior of the building was enclosed with a roofed veranda and embellished with a grilled iron fencing. The glistening, white walls and green shutters, made a pretentious picture of conservative simplicity.

The main floor was divided through the center by a large drawing room. At one side, the main entry hall gave access to a library and a small family parlor. On the other side of the drawing room was located the dining area which was separated from the pantry by a hallway leading to a side entrance. A dumbwaiter was installed in the pantry to serve the upper floor. The entire building was heated by the basement furnace which circulated heat to each room.

From the main entry hall, a stairway led to the upper floor which consisted of eight bedrooms and a central bathroom. The four corner rooms were equipped with fireplaces. Each bedroom was furnished with a heavy highboard wooden bed, wardrobe, dresser and a basin and pitcher on a commode. The walls displayed colorful wallpaper in naturalistic motifs in harmony with full-floor carpeting. Period furniture prevailed throughout the building. The general decor and furnishings of the interior were a true expression of gracious living.

In 1865, J. Ross Browne visited New Almaden, where he spent sometime in making a general report of the quicksilver mines for the United States government. The following is an excerpt from an article in Harper's Monthly, October 1865:

"From San Jose to Hacienda, a distance of twelve miles, the country is nearly level and the road is fringed with luxuriant groves of cottonwood, sycamore and willow. A more delightful drive is not to be found in California. Two lines of stages make a daily trip each way; so that passengers from San Francisco are enabled to reach the mines within four to five hours. Private conveyances are always to be had, especially if the visitor be so fortunate as to enjoy the acquaintance of the general agent and factor, whose kindness and hospitality are proverbial.

Entering the shady groves of the Arroyo de los Alamitos, the road winds along the declivities of the canyon for a distance of a mile and a half, when a little to the left is seen the capacious mansion known as the Hacienda, (Casa Grande) or headquarters, its massive walls and broad verandas embosomed in shrubbery. The eye delights to rest upon such a scene of rural comfort, which presents not only the rarest natural beauties but the highest evidences of cultivated taste. Flowers of rich and variegated hues bloom in the gardens, Graveled walks traverse the orchards and graperies, and wind through the umbrageous groves that fringe the Alamitos. The music of the rippling waters chimes pleasantly with the singing of the birds and the happy voices of children. An air of luxury and refinement pervades the premises. All the office and appointments are in excellent taste, combining simplicity and rural effects with convenience and elegance. One is reminded of the sumptuous and old-fashioned homes in the South.

Here, all is peace and prosperity. Carriages are drawn up before the door; spirited horses, saddled for the ladies and their attendants, prance impatiently at their posts; easy-chairs scattered along the veranda. Happy is he who can spare a few days from his business in the city and partake of the social amenities that prevail at the Hacienda; for here is luxury without pretense, elegance without restraint and hospitality without affectation, where all may enjoy that rarest privilege—perfect liberty to be happy in their own way."

The first occupant of the Casa Grande was mine manager, Henry Halleck, who remained until 1863. With the change of mine ownership, Samuel Butterworth took residence until 1870. The longest resident was James Randol until 1892. With the departure of Randol, his successor, Robert Bulmore became the last official resident until 1900. Following 1900, the building came under private ownership and passed away the years until the property was purchased for commercial reasons.

One of the features that attracted much notice, was the discovery of a mineral spring on the banks of the Los Alamitos creek. In December, 1867, the Quicksilver Mining Company leased 2.5 acres to F. A. L. Pioche for a period of ten years for the sale of what was to be known as The New Almaden Vichy Water. The elixir of life and cure-all for every known affliction was distributed by the California Vichy Water Company from their depot in San Francisco. The product was given national advertising and one seeking its attributes could get 12 bottles for four dollars.

In expressing to the public the merits of the product, the following statement appeared in newspapers and magazine advertisements:

"This Mineral Water is highly efficacious in cases of: Impoverishment of the blood—Weakness of the nervous system—Chronic Inflammation of the Liver, Stomach, Spleen and Inflammation generally, (when there is no fever)—Dyspepsia in all its stages—Loss of appetite—Bad Digestion—Constant and tenacious vomiting—Obstruction of Liver and Spleen—Night sweats—Chronic rheumatism, (when without fever)—Gout, when the patient is still able to walk—Gravel, and in cases of Diabetes, the New Almaden Vichy Water is a most powerful adjuvant to remedies prescribed for this terrible disease.

This Mineral Water is a most agreeable beverage, pure or mxied with water, Beef Broth, Red Wine, White Wine, Brandy, Sherry Syrup, Champagne, Milk, etc.

This Water restores lost strength, energy and good digestion to the stomach ruined by excess eating and drinking, immoderate smoking and chewing, excess of work or pleasure; it restores strength to the stomach weakened by excess of labor, sedentary occupation, lack of out-door exercise, etc. it corrects the debility caused by certain medical treatments, that have either been pursued too long or that have been badly managed; many chronic diseases, especially those of the abdominal organs, are alleviated by the use of this Water.

This Mineral Water is especially useful and very powerful not only in the course of certain diseases, but also as a preventive of intermittent fever, chronic rheumatism, scrofula, general debility arising from disorder of the nervous system and circulation of the blood, and especially of the venous circulation."

The variety of disorders, selected by the producers which appeared to be generally common with most people, created a lucrative market for this unusual product. Gallons of mineral water were shipped from the Hacienda to San Francisco where the bottled product found a receptive market.

The one remaining evidence of the New Almaden days, is the small, unpretentious village, once known as the Hacienda. Today, this area is referred to as New Almaden and life goes on in the same secluded manner. For many years, following the liquidation of the Quicksilver Mining Company, the little settlement was practically deserted as the inhabitants moved out to seek new fields of endeavor. The Hacienda, like the neighboring settlements on the Hill, took on the atmosphere of a ghost town. However, life was perpetuated for the Hacienda as newcomers, discovering the quaintness of this isolated setting of empty cottages, moved in as permanent or vacation residents. Little has changed along the shaded thoroughfare. The original cottages of adobe and wood, have been altered and renovated and with the addition of a small number of new structures, the setting still conveys a certain character that is reminiscent of its former years.

The spirit of New Almaden still prevails and is colorfully expressed with an annual parade and jamboree held in September to commemorate the "Old Almaden Days." This affair and other civic functions are sponsored by community organizations. At several important locations about the village, the New Almaden Historical Society has placed markers denoting the historical significance of the subject. The old adobe store, standing since 1849, is still in service with a bar room and small

THE CINNABAR FLOAT

This float drawn by four white horses, participated in the Carnival of Roses parade held in San Jose in May, 1901. The colorful affair was to honor the visit of President William McKinley and his wife to the Santa Clara Valley. The entry was sponsored by the Quicksilver Mining Company.

THE HACIENDA OFFICE STAFF

This interior was taken in 1902. The three gentlemen in charge of the company's business at the Furnace Yard are: Left — M. H. Harms, Assistant Bookkeeper. Center — George Carson, Cashier. Right — A. Wiener, Accountant.

grocery store. The original adobe houses have been artistically renovated for residences and a museum. The St. Anthony's Roman Catholic church is still in a good state of preservation and has been in religious service since 1900. At the crossing of the creek stands the Cafe del Rio, a building remodeled from the once Hacienda hotel and serves a popular cuisine to its many out-of-town customers. The Casa Grande still stands, sturdy and intact as the day it was built. The original title no longer exists and its identification with the glory days of New Almaden, has long faded into the realm of legend. It is now known to the passersby, as Club Almaden and for some years has been a private recreational center for the public. The once spacious and colorful gardens have become a car parking area and the picturesque lake has been converted into standard swimming pools. The old structure still retains an attractiveness and for the uninformed, a roadside marker states its true identity. Adjoining the resort property, a high picket stockade encloses an area called "The Theater Under the Stars." During the summer season, an amateur group known as the "Wagon Stagers," offers productions with a flavor of the old mining days. A U. S. Forestry firehouse stands on the site that was once a small park where the Brass Band used to perform on occasions.

The days of the Hacienda have become legendary and only the "old-timers" think of the settlement by this title. For the present residents, it is New Almaden and a casual conversation along the street will bring out the comment, that this was once a great mining town.

Furnaces 1 and 2 located on the banks of the Alamitos creek. These furnaces designed by H. J. Huttner, engineer, and built by Robert Scott, brick mason, revolutionized the reduction of quicksilver by efficiency to maintain continuous firing with greater economy and recovery of the liquid metal. *(Winn Photo)*

A rear view of the Reduction Works with the tramway on the back hill. Photo 1886. *(Winn Photo)*

A view of the Reduction Works as it appeared in 1885. On the hill to the right, the tramway descended from the mines to transport the ore. As the furnaces roasted the ore, the mercurial fumes were passed off up a brick flume to a chimney outlet on the hill. As a landmark to this setting, the chimney is still standing today. The Reduction works was dismantled during the years following the end of operation by the Quicksilver Mining Company in 1912.

FLASKING QUICKSILVER

This photo taken at the Reduction Works in Hacienda, shows the procedure of preparing quicksilver for shipment. These iron flasks were imported from Spain and held 76½ pounds. This odd weight was derived from the Spanish measurement by which the flasks contained 75 "libras" or 1.0143 pounds. In later years the half pound was dropped and the flasks weighed an even 76 pounds. Each flask was filled and weighed to meet exact specifications. The man shown engaged in this activity is Henry Tregoning.

THE HACIENDA DE BENEFICIO — 1875

During the early mining days of the Barron, Forbes Company, the furnace plant was known as the Hacienda de Beneficio, which meant Reduction Works. The little settlement at the approach to the furnace yard became identified as The Hacienda. Through the years of mining activity, over 1,000,000 flasks of quicksilver were produced here. Great piles of firewood were continuously stacked to supply the constant demand of the furnaces. *(Watkins Photo)*

TRANSPORTATION BY STAGE COACH

With a full load, the stage is about to start the 12 mile journey to San Jose. A daily stage left in the morning and returned to the Hacienda in the late afternoon. Pancho Alcaraz is the driver. Photo taken in 1896.

THE ROCKAWAY

The Almadener was well served for transportation to the railway depot or round-trip to San Jose. This vehicle of three seats and a fringed top was first class for the day.

AMATEUR THEATRICALS ENTERTAINED THE VILLAGERS
This photo taken in August, 1892, portrays the cast for the production, "Among The Breakers," which was presented at the Helping Hand Hall in Hacienda. The cast for this picture assembled on the steps of the Casa Grande. Garbed in their characterizations are:
James Harry, Beatrice Hall, J. Wilkinson, I. Mendizabal, Kate Hall, George Carson, Lottie Bulmore,
Robert Bulmore, Mrs. Robert Bulmore and Charles Derby.

THERE WAS MUSIC IN THOSE ALMADEN HILLS
The villagers were entertained at various community functions by the colorful costumed brass band. This small but capable group of musicians played a popular role in the life of this mining settlement. Front Row, L to R — Arnold Vineent, Juan Paredi, Cui Mercado, Andres Sambrano, Henry Vincent, Dan Flanagan. Middle row, L to R — Unidentified, Antonio Parades, Feliciano Martinez, Juan Mattos, Joe Varrote. Top row, L to R — Adolph Martinez, Amado Gonzales, John Luxon, Abran Martinez.

The baseball team of 1912 was experiencing less support from the Almadeners as the mines were closing down and the population began to move out.

Nine players and a mascot composed the first uniformed baseball team in 1894. Games were played at the Hacienda school grounds with visiting teams from surrounding townships.

CHAPTER 8

CONCLUSION

As the new century approached, borrasca was slowly descending upon the working of Mine Hill. Borrasca was a common term in many mining camps and signified the reckoning day when pay-dirt became harder to find. The rich ore of the Cinnabar Hills had run its course. By 1899, ore production was below the tonnage for continuing with any future promise of success. Drastic economy measures were applied with almost a complete reduction of the employees. The position of General Agent was abolished, terminating the services of Robert Bulmore. Charles Derby was put in charge of the entire operation which he carried on until 1901. When he resigned, he was replaced by his father, Thomas Derby, who remained until 1909.

While Randol was promoting the fruitful years of Mine Hill, he was never too receptive to suggestions or recommendations from his associates in regard to developing the outside mines along the ridge to the West. These mines had been established during the early years of Barron, Forbes operation. Work was concentrated on opening the Providencia, San Mateo, Senator, San Antonio and Enriquita shafts but development work of any merit was curtailed for lack of capital.

With the resignation of Thomas Derby, E. J. Furst was engaged in a desperate move to keep the mining activity alive. He made a survey of the existing prospects and decided to concentrate all of the activity on the old Senator mine which appeared the most favorable. John Drew, a veteran in quicksilver mining, was placed in charge and a new Senator shaft was constructed during the year 1909. During the preliminary stages of operation, Furst resigned and was replaced by R. Nones, who engaged the services of J. F. Tathum as superintendent. A fair degree of success was achieved as new ore bodies were uncovered but the amount of good ore was insufficient for continuing further operation. By 1912 operations had resorted to working the dumps and debris which had been discarded during the glory years, but because of poor financial management little profit was realized. In these closing hours, a skeletal crew made futile efforts to sustain the productive life of the New Almaden mines but the results were unrewarding. The Quicksilver Mining Company had reached the end of the road for there was little left that would warrant its existence. By the close of 1912, the Cinnabar Hills could no longer support further operations and the company closed their books and officially declared the entire holdings in a state of bankruptcy.

The years that followed consisted of sporadic ventures in recovering the remaining fragments of ore from waste and discard. In 1915, George Sexton obtained a 25 year lease and engaged W. H. Landers as general manager. The Senator mine was again prospected with John Drew in charge. As World War I assumed greater proportions, quicksilver became an important item and the price of $81.52 a flask brought rejuvenated efforts to find more quicksilver. A flask selling for $49.05 in 1914 increased to $114.03 in 1918. During these war years, 9,073 flasks of quicksilver were recovered for a gross income of $873,529.

Actual mining activity, again, came to an end with the closing of the Senator mine in March, 1926. The death of George Sexton, president and leading stockholder, stopped the activity and the property entered another period of litigation. From 1928 to the present day, there has been continuous activity by small operators in retorting the last gleanings of ore available, and retrieving lost quicksilver from previous operations throughout the furnace yard in the Hacienda.

In 1935, C. N. Shuette, a specialist in quicksilver mining, was engaged by the property owners to evaluate the existing conditions and the potential for further speculation. His report gave indications of promise and on this basis, the property was leased by the W. H. Newbold Company of Philadelphia. Operations commenced in May 1940 under a newly formed group known as the New Almaden Corporation with C. N. Shuette as general manager.

The Cinnabar Hills, again, partially came to life and continued through World War II with an estimated production of 7,000 flasks. Heavy equipment cut through the dumps and sliced away the steep slopes of Mine Hill. The results of the open-cut operation were favorable and in 1940 a reduction plant with a 100 ton rotary furnace was constructed. Work continued until December 1945, when the prevailing conditions offered little promise for de-

THE LAST DAYS OF THE SENATOR MINE

Located several miles northwest of Mine Hill was the Senator mine which was the most productive of the workings outside of the main mining area. The mine had brief periods of activity from 1872 until 1900. In 1909, the Quicksilver Mining Company made a concerted effort to stay in business by reopening operations with John Drew in charge. Work was terminated in 1912, when the Company was declared bankrupt. Operations were resumed again in 1915 by George Sexton who took a 25 year lease on the New Almaden property. Mining facilities were extended and a reduction plant was installed. Mining continued with intervals of inactivity until 1926 at which time the mine was closed and the reduction plant dismantled. The last years of activity had produced some 20,000 flasks of quicksilver.

(Courtesy John Gordon)

This is the sight as it appeared in 1924, which contained the furnaces where millions of tons of cinnabar were roasted, during a half century. The structures were dismantled soon after the termination of activity by the Quicksilver Mining Company. In the background, the last remnant standing is the Company office, while to the right is the settlement of the Hacienda.

(Courtesy John Gordon)

HERE, THEY FILLED THE FLASKS WITH QUICKSILVER

This is the site of the once great reduction works where the furnaces of Hacienda handled the ore from the New Almaden mines to create world-wide prominence for its quicksilver production. The buildings have disappeared and the ground has been excavated to regain losses experienced during the early day processes. Ascending the hill, can be seen the brick constructed flume which carried the escaping vapors to the outlet through two, tall brick chimneys on the hill.   (Courtesy John Gordon)

veloping a large scale enterprize. Once again, a company organized operation came to a close. The equipment was liquidated and the company was dissolved.

However, all is not quiet on the slopes of Mine Hill. A small group of die-hards still continues the search and small scale production survives from the fragments of sparsely distributed ore, in and about the old mine shafts of yesteryears.

The life and times in the Cinnabar Hills of New Almaden have faded into obscurity and have become almost legendary since the last miner, James Prout and his wife, left the ghostly setting of empty buildings on the Hill in 1912. With over a half century of quicksilver production, the Cinnabar Hills yielded of its treasure to share in the turbulent and romantic era of California's golden years. The tunnels and shafts, in their subterranean complexity are dark and silent and the seepage of water has filled the cavernous labores and deteriorated the timbers. No longers does the sound of explosives reverberate through the deep workings and quiver the surface of Mine Hill or rattle the windows of the miner's homes. The voices of the Cornish and Mexican miners are but echoes, entombed with the sound of grinding drills and the grating of the ore cars. Water continues to fill the excavations of man, for the great Buena Vista pumps have stopped. The ore cars down the incline have made their last descent, for the rich pay loads of Cinnabar have run out. The mighty furnaces, that roasted millions of tons of ore and produced over a million flasks of quicksilver, have long since been removed from the scene. High on the western ridge, the tall, brick chimneys which served their function as vapor outlets, have become sentinels against the skyline, as monuments to the bonanza days of a quicksilver mine, unprecendented in its prestige and contribution. The shaft structures of the Randol, Buena Vista, Santa Isabel and others have long been dismantled and rubble obliterates their identity.

A half century has passed since the blast of the shaft whistle was heard, the ringing of the schoolhouse bell and the voices of the people have echoed through Deep Gulch. The tinkling bells of the six horse teams pulling their cargo up the curving two-mile road, will no longer alert the hill-top people of the busy thoroughfare. Along the foot-paths, the trudging, weary, miners have left their last footprints in the red, dusty soil. The bonanza years have faded silently into the stark realism that typifies the life of every mining camp. It is like the seed in the field that matures to blossom forth with all its inceptive qualities, only to eventually wither away to conclude the cycle of its existence.

The cave of the red rock that offered Mohetka to the native Indians and the ghostly figure of Andres Castillero, have become a legend. The quicksilver days of New Almaden will exist only in the reminiscent thoughts of the remaining pioneers and their descendants. For the many, who loaded their belongings and departed down the road to a new beginning, their life on the Hill left a lasting impression. Each had been an integral part of a community in which the neighborly spirit had been a prevailing custom. The day, when the shafts would become silent and the little white houses, with colorful gardens would be vacated, was a picture, not conceived by the devout Almadener, for they had shared in an environment to which they anticipated no end. The ore was inexhaustible, so they were told and the Hill would survive far beyond their lifetime.

The lively settlement of Spanishtown and the conservative Englishtown, became deserted and unattended. For some twenty years, the company buildings and miner's homes, endured a period of deterioration. The elements of nature methodically transformed the hilltop domain into a setting of lifeless survival. The Hill portrayed the picture, typical of many deserted mining camps and through its remaining years, slumbered away in the true essence of ghost town traditions. Occasionally, old Almadeners, would return in a nostalgic mood, to look at a setting of many memories and particularly to see the fast, crumbling cottage that once was called home.

An article appeared in the San Jose Mercury, December 21, 1912 which gives a brief picture of the New Almaden exodus.

ABANDONED HOMES TO FOLLOW CINNABAR

"Cinnabar, playing the role of the Pied Piper, has made a modern Hamlin Town of the Hill village of New Almaden. In this instance, however, not only the children ran, but the entire population, has heeded the Piper's call and home, schoolhouse and store are closed and abandoned to the rats.

The inhabitants of the Hill village, including about 150 children, following the paying lode of the largest quicksilver mine in the world, are now housed in the Hacienda, three miles from their old town on the mountaintop, in picturesque adobes constructed in the romantic past.

A petition for the consolidation of the Hacienda and Hill school is now being circulated in both districts and some action is

CHINESE SAM

One of the popular employees about the mining property was Chinese Sam. When Sam died in February, 1889, a subscription was taken by the miners amounting to $90 and the company gave an additional $110. The donation was sent to the Chinese Consul in San Francisco for the purpose of transporting Sam's family back to China. The company also assumed the costs of Sam's funeral.

(Winn Photo)

THE ACEQUIA ALONG THE THOROUGHFARE
IN 1908

The Spanish term for the man-made ditch that carried water from the Reduction Works through the Hacienda, was Acequia. It offered water for the gardens and also a playground for the children. In the background may be seen Bohlman's Livery Stable which was the headquarters for all stage service.

(Courtesy of Alice Hare)

expected at the meeting of the Board of Supervisors.

A flourishing town once existed on the mountaintop. In the hey-day of its glory, four teachers were employed and from 150 to 200 children attended its schools."

With the passing years, the quicksilver days of New Almaden have become further removed from any present day reality. The riches of the Cinnabar Hills and the life and times that covered a half century, have almost completely disappeared with the gradual passing of its pioneers. Little was re-corded that would continue its prestige as one of California's famous mining camps. However, the settlement of New Almaden will continue to exist, if only in name, and will be perpetuated for its significance with the early history of California's great mining era. After many forgotten years, recognition has come to the last remnants of a quicksilver mine by the acknowledgement of the federal government. On August 2, 1964, a plaque was unveiled by officials of the U. S. Department of Interior, which now includes New Almaden in the registry of national historical landmarks.

THE HACIENDA HOTEL IN 1924

Originally a boarding house, the building was converted into a small hotel to accommodate visitors at the mining settlement. After the departure of the population, the building stood vacant for many years until it was remodeled to serve as a restaurant. Today, the old structure with new trim and color, continues in business as the Cafe Del Rio. *(Courtesy John Gordon)*

A GHOSTLY REMNANT OF THE QUICKSILVER DAYS

Stillness descended upon The Hacienda after the mining days officially closed in 1912. Only a few hangers-on remained in the village of deserted buildings that portrayed a ghostly facade along the main thoroughfare. This scene shows two vacated adobe structures as they appeared in 1924. In later years, the quaint setting was discovered and a new population began moving in to establish residence as permanent and weekend homes. Most of the buildings, with alterations are in service today.

(Courtesy John Gordon)

A DESERTED ADOBE IN 1924

This is one of several adobes built in the 1850's that weathered the quiet ghost-like years to continue its existence with new trim and furnishings.

THE HELPING HAND CLUB IN THE HACIENDA — 1924

After 1912, the building that served for many social functions of the mining settlements was deserted and waited out its days until a wrecking crew salvaged its material value.
(*Courtesy John Gordon*)

Casa Grande gardens with pagoda. In the background is the Hacienda school house.

A TEAHOUSE FROM CHINA

This authentic pagoda was shipped to New Almaden by the Emperor of China. It was assembled in the spacious gardens of the Casa Grande and became a novel attraction for the many visitors. This gratuitous gesture was in acknowledgment of a colorful reception extended the Chinese emissary on his visit to the mines in the 1850's.
(Courtesy New Almaden Historical Society)

117

The Chinese pagoda, fading with neglect, as it appeared in 1924.

In 1874, **The Hacienda** experienced a fire which destroyed many of the miner's homes. Soon after, James Randol installed a system of fire hydrants and constructed a bell tower with hose cart. As the years passed this structure served on many occasions to alert the village and call the bucket brigade to action. This is the scene in 1924 when the thoroughfare was quiet with empty cottages. *(Courtesy John Gordon)*

The Almadeners expressed themselves for the special
occasion of commemorating the Centennial of George
Washington's Inauguration to the Presidency of
the United States.

Stacks of quicksilver flasks at the office yard in the Hacienda. One of the tramway inclines can be seen on the hill.

Recent Scenes of New Almaden
—MILTON LANYON

Modern equipment cuts away the rocky
slopes of Mine Hill.

This 100-ton rotary furnace was constructed in 1940 to
roast the ore obtained from open-cut mining. The present
day value of quicksilver at $500 a flask, encourages the
search for ore which keeps the furnace in operation.

The Powder House near the site of Englishtown is the
only remaining structure that has weathered the years.

The Mine Office as it appeared in its last days.

Site of the first mining operation in California.

St. Anthony's Catholic Church built in 1900.

One of the original adobe houses.

A lone marker in the vine shrouded cemetery in what was originally The Hacienda.

The Casa Grande continues in service as a residence, post office and public accommodations. This once pretentious dwelling and headquarters for the mine managers is now Club Almaden.

Along the shaded thoroughfare still stand the wooden and adobe dwellings of the quicksilver days.

A typical cottage that was home for the miner in the several settlements of New Almaden.

Former site of the Catholic Church on the
ridge above Spanishtown.

Deep Gulch that contained the settlement of Spanishtown.

Last remnants of the old store in Englishtown.

New Almaden Day is celebrated annually with parade,
barbecue and dancing. The affair is sponsored by the
local citizens.

The original New Almaden store building.

The Cafe Del Rio on the banks of the Alamitos Creek.

Old adobe which serves as residence and museum.

Spanish Terms Commonly Used In New Almaden

1.	ACEQUIA	ah-thay-ke-ah	A canal, trench or drain
2.	ADIOS	ah-de-ohs	Good-by
3.	ADOBE	ah-do-bay	Bricks made from earth, re-enforced with straw and sun-dried
4.	ALTA	ahl-tah	A hanging wall in a mine
5.	BOLETA	bo-lay-tah	Voucher or warrant for receiving money or other things
6.	CARGA	car-gah	Load, burden, freight
7.	CAMINO	cah-mee-no	Beaten road
8.	CASCARONE	cahs-cah-ro-nay	An egg shell filled with colored paper
9.	COLGANTE	col-gahn-tay	Act of hanging, clinging
10.	DINERO	de-nay-ro	Coin, money, gold, coinage
11.	ESCALERO	ays-cah-lay-ro	A ladder, stair
12.	GRANZA	grahn-za	Ore mixed with other rock or substance
13.	GRUESSO	groo-ay-so	Purest quality of selected ore
14.	HACIENDA	ah-thee-en-da	Pertaining to the Reduction works
15.	HOMBRE	ohm-bray	Referring to a man
16.	HILOS	eel-yohs	Small veins of Cinnabar
17.	LABORE	la-bo-ray	Ore stopes in the mines
18.	MINERO	me-nay-ro	Miner who digs for metals
19.	MERCURIO	mer-coo-re-o	Quicksilver
20.	MISA DE GALLO	mee-sah day-gahl-lyo	Early mass, mass of the rooster
21.	NOCHE BUENA	no-chay boo-ay-nah	Pertaining to Christmas Eve
22.	PANADERO	pah-nan-day-ro	Baker, seller of bread
23.	PINATA	pee-nah-tah	Clay or paper receptacle
24.	PLANILLA	plan-neel-lyah	Term applied to ore shed
25.	POSADA	po-sah-da	An inn, tavern, lodging house
26.	REBAJO	ray-bah-ho	Cutting of a groove in stone or hillside
27.	TALEGOS	tah-lay-gos	Bag or sack
28.	TANATE	tah-nah-tay	Leather bag or sack
29.	TANATERO	tay-nah-tay-ro	Term applied to laborers who carried ore
30.	TIERRA	tee-ayr-rah	Inferior ore, found in refuse earth
31.	VARA	vah-ra	Spanish measure of length

REFERENCES

Bailey, Edgar H. and Donald L. Everhart, GEOLOGY and QUICKSILVER DEPOSITS of the NEW AL-
MADEN DISTRICT, Division of Mines and Geology, State of California, 1964.

Barron, Forbes Company, California Historical Society, June 1936.

Browne, J. Ross, DOWN IN THE CINNABAR MINES, Harper's New Monthly, October 1865, Vol. XXXV.

Downer, Mrs. S. A. THE QUICKSILVER MINES OF NEW ALMADEN, California Monthly Magazine,
October 1854, Vol. 2.

Foote, Mary Halleck, A CALIFORNIA MINING CAMP, Scribner's Monthly, October 1865, Vol. XXV.
Lakes, Arthur, THE NEW ALMADEN MINES, Mines and Minerals, March 1899, Vol. XIX No. 8.
Leicester, Henry M. THE FIRST CHEMICAL INDUSTRY IN CALIFORNIA, Journal Chemical Educa-
tion, November 1943, Vol. XX.

Lord, John Keast. A TRIP TO THE ALMADEN MINES, 1860. Bancroft Library, Berkeley, California.

Lyman, Chester S. AROUND THE HORN TO THE SANDWICH ISLANDS AND CALIFORNIA, 1845-
1850.

ON MINES OF CINNABAR IN UPPER CALIFORNIA, American Journal of Science and Arts,
November 1848, Vol. VI., University of California Library, Berkeley, California.

Munroe-Fraser, J. P. THE HISTORY OF SANTA CLARA COUNTY, Alley, Bowers and Company, San
Francisco, 1881.

Randol, James B. REPORT ON MINERAL INDUSTRIES IN UNITED STATES, Eleventh Census, 1890.
Government Printing Office, Washington, D.C.

Shutes, Milton H. ABRAHAM LINCOLN AND THE NEW ALMADEN MINES, California Historical
Society Quarterly, March 1936, Vol. XV.

Taylor, Frank J. THE CALIFORNIA MAGAZINE OF PACIFIC BUSINESS, March, 1937.

Wells, William V. THE QUICKSILVER MINES OF NEW ALMADEN, Harper's New Monthly, 1863,
Vol. XXVII.

THE AUTHORS

MILTON LANYON is a native of San Jose, California and was graduated from San Jose State College and Stanford University. He served on the faculty at San Jose State College for 23 years in the departments of Art and Education and has recently retired as Professor Emeritus. His father left Truro, Cornwall as a young man for the mining areas of the West and finally, settled in New Almaden in 1878, where he was engaged as a miner, shift boss and a short term as constable of the New Almaden Township.

LAURENCE BULMORE is a native of New Almaden. As a youth he experienced the twilight years of the quicksilver days. His father, Robert Bulmore, was the last manager for the Quicksilver Mining Company and last official resident of the Casa Grande. The acquisition of mining data and the rare collection of photographs taken by his father, establishes Laurence Bulmore as an authoritative source of general information of the New Almaden days. Before retirement, Mr. Bulmore served for many years as an engineer on the San Francisco Bay ferry boats.

Book Design and Art Work
MILTON LANYON

Text and Photo Arrangement
ROBERT WYATT

PRINTED BY THE VILLAGE PRINTERS
LOS GATOS, CALIFORNIA, U.S.A.